The Paper Machine Series Volume I

THE WORKING CLOCK/TIMER

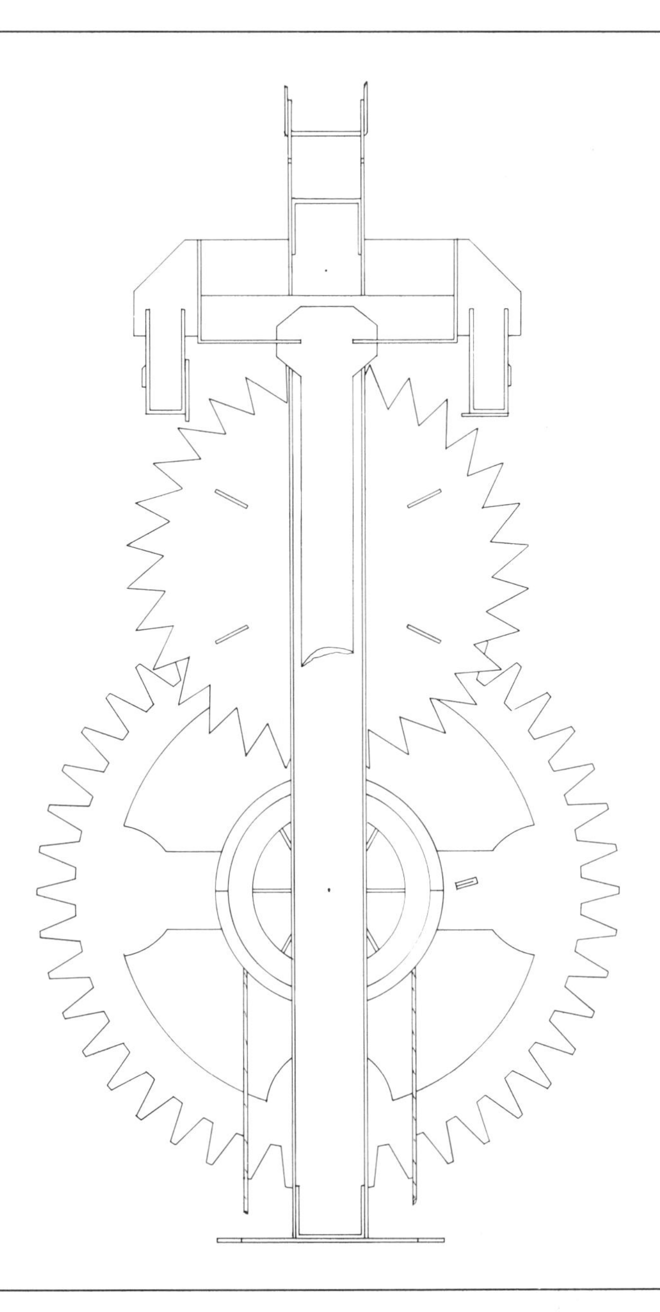

JOEL MOSKOWITZ

Simon and Schuster • New York

 Published by Simon and Schuster, a division of Simon & Schuster, Inc., Simon & Schuster Building, Rockefeller Center, 1230 Avenue of the Americas, New York, New York 10020. SIMON AND SCHUSTER and colophon are registered trademarks of Simon & Schuster, Inc.

Designed by Barbara M. Marks

Manufactured in the United States of America

1 3 5 7 9 10 8 6 4 2

Library of Congress Cataloging in Publication Data

Moskowitz, Joel.
The working clock/timer.

(The Paper machine series; v. 1)
Bibliography: p.
1.Clock and watch making. 2.Paper work.
I.Title. II.Series.
TS548.M67 1985 681.1'13 85-14414

ISBN: 0-671-55183-3

ACKNOWLEDGMENTS

There is a tremendous amount of work connected with taking an idea and turning it into something you can hold in your hand. Many people encouraged me, did a lot of work, and pushed the project along. I would like to thank all of these people. Without them there would be nothing.

Ester Moskowitz
Tracey M. Siesser
James E. Korenthal
Bonnie Schertz
Meredith Bernstein
John Herman
Jacquie Saul
Howard Goldstein
Steve Konopka
Joseph W. Smith III
Ron Sullivan

A NOTE ON METRIC UNITS

This project was designed using English units of measurement. If you normally use metric units, the following conversion figures are included for your convenience:

1-inch pins = 25.4 mm (anything over 20 mm is fine)

7 feet of string = 2.1 m (use about 2.15 m so you have 2 m for the weights and 150 mm to make the loop to attach the timer to the wall)

The weights are given in terms of U.S. pennies, which are copper disks about 19 mm in diameter and 2 mm thick. They weigh roughly 3.1 gms each. As long as the total weight that hangs down is enough to power the timer and not enough to pull the clock apart, anything close will work. You can even use sand.

CONTENTS

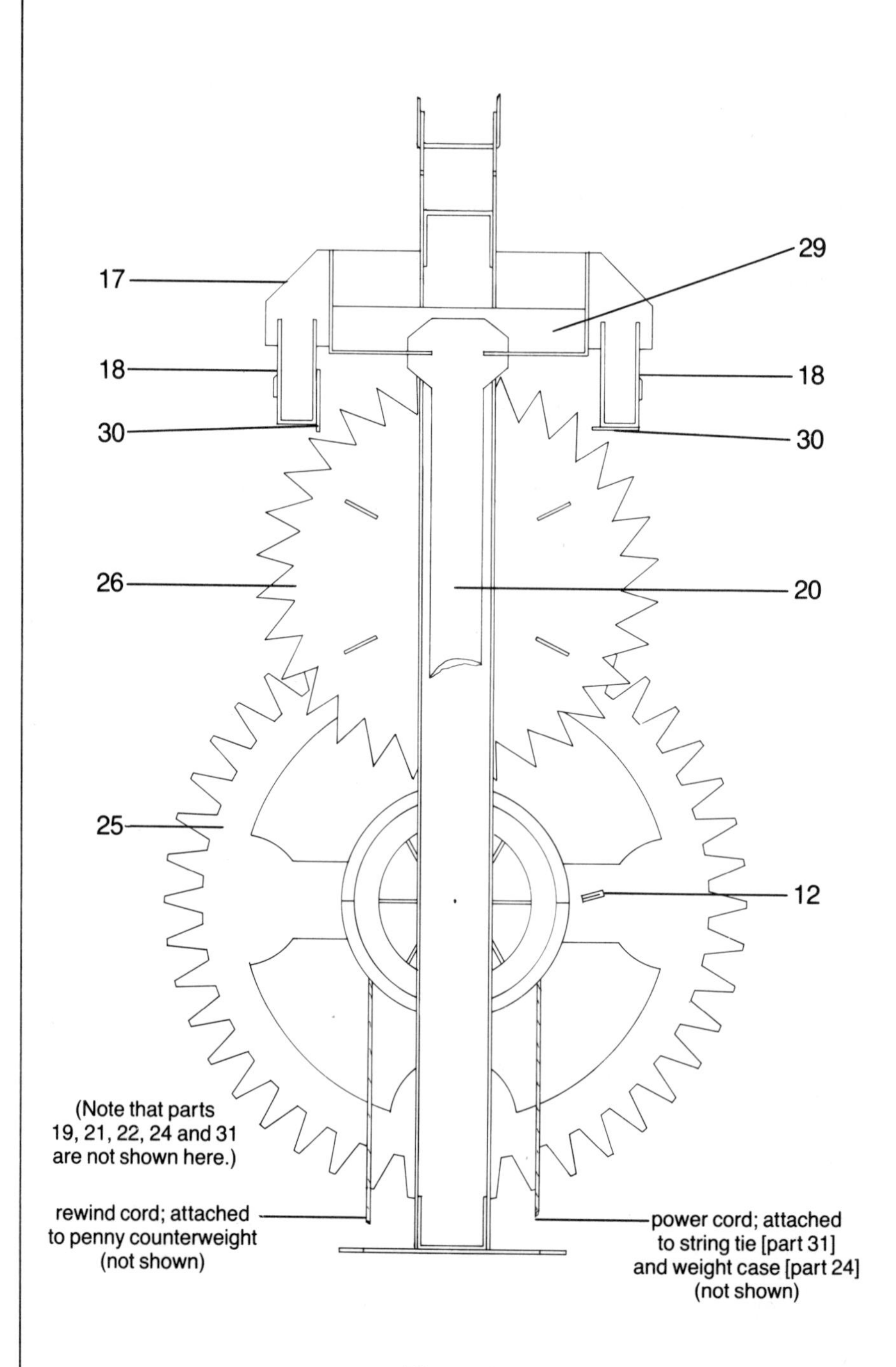
29
17
18
18
30
30
26
20
25
12
(Note that parts
19, 21, 22, 24 and 31
are not shown here.)
rewind cord; attached
to penny counterweight
(not shown)
power cord; attached
to string tie [part 31]
and weight case [part 24]
(not shown)

Figure 1

BILL OF MATERIALS

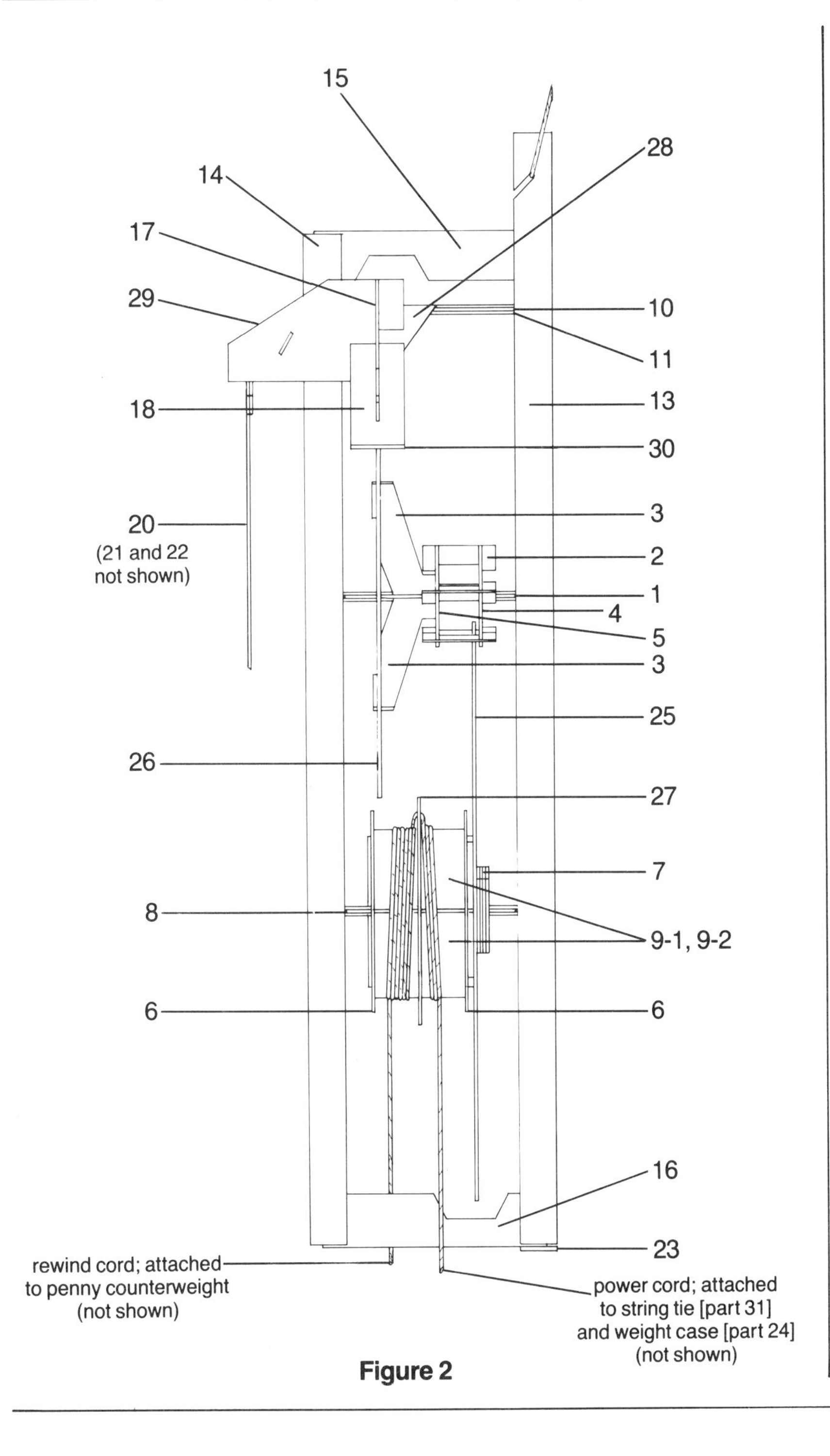

Figure 2

1 Escape Wheel Arbor Core
2-1, 2, 3, 4, 5, 6 Pinion Wheel Teeth
3-1, 2 Escape Wheel Arbor
4 Outer Pinion Wheel
5 Inner Pinon Wheel
6-1, 2 Winding Drum Ring
7-1, 2, 3, 4 Ratchet Wheel
8 Great Wheel Arbor Core
9-1, 2 Great Wheel Arbor
10 Pallet Arbor Core
11 Pallet Arbor Bottom
12-1, 2 Ratchet Click
13 Rear Frame
14 Front Frame
15 Top Frame Spacer
16 Bottom Frame Spacer
17 Pallet Arm
18-1, 2 Pallet Tooth
19 Pendulum Weight Case
20 Top Pendulum Rod
21 Middle Pendulum Rod
22 Bottom Pendulum Rod
23 Stabilizer
24 Weight Case
25 Great Wheel
26 Escape Wheel
27-1, 2 Winding Drum Spacer
28 Pallet Arbor Top
29 Crutch
30-1, 2 Pallet Shims
31 String Tie
32 String—Approximately 7 feet (not included)
33 22 Pennies (not included)
34 6 Straight Pins (not included)

LIST OF ILLUSTRATIONS

INTRODUCTION

Mechanical clocks, with moving gears and swinging weights, have been the source of endless fascination for seven centuries. Here is your chance to get in on the fun and build your own working clock-timer.

The basic mechanism of a clock is simple and building one is a question of what tools are at hand. In my case the answer was "None!" so I resorted to paper, which I could cut and form with a knife, and some glue.

All the parts of the model, including the gears, are included on the three pages of heavy paper that are bound in at the back of this book. The parts are precut and prescored for folding, which eliminates a lot of the tedious work you would normally have to do in most paper models. The only additional materials you need are several straight pins, string, two dozen pennies, white glue, and masking tape. The only tools you need are a sharp knife and a straightedge. As long as you exercise a little care you should have no trouble building an accurate timer.

The biggest cause for failure in modelmaking is not reading the instructions beforehand. It is essential that you read and understand the instructions in this book before you begin. That way, you will be able to visualize how everything fits together and not make careless errors. In my own case, I used to plunge ahead without any patience for little details like instructions. I usually figured out the gist of the project, but I lost out on the subtle details, and the end product reflected that.

How the Timer Works

The brain of any clock is the escapement, which regulates the speed of the clock and keeps it accurate. The muscles of the clock are the weights, which provide the power to turn the gears. The weights and escapement work in tandem. The weights tug down, trying to make the gears spin as fast as possible, and the escapement alternately blocks and then releases the timer gears from turning, so the average speed of the clock is just on time. You can picture this very easily if you think of a gear pulled around by a string wrapped around it, like a yo-yo going down. A weight pulling down at the end of the string would make the gear spin very fast. By sticking a rod between the teeth of the gear you could stop the wheel from turning. Then by quickly pulling the rod in and out between the teeth you could regulate the average speed of the gear. That is how an escapement works. Instead of a normal gear, an escape gear, which has specially shaped teeth, is used. Instead of a wooden stick, the escape teeth are blocked by hard blocks of material called pallets that are mounted on a pallet arbor. The pallet arbor looks like a wishbone with a pallet on each end, and it pivots, tilting back and forth, alternately blocking an escape wheel tooth on each side of the wheel.

The trick is to block and then free the escape wheel at the correct intervals. Over the centuries a whole host of methods were developed, but the model in this kit uses one of the best and certainly the most popular. The intervals are determined by a swinging pendulum which is connected to the pallet arbor. A swinging pendulum will always take the same amount of time to swing from side to side, no matter how heavy the pendulum is or how wide the particular swing is. The duration of the swing is directly related to the length of the pendulum, nothing else. (This isn't entirely true, but as long as the angle of swing is small, the differences are not detectable except with special equipment.) Each time the pendulum swings up, freeing a tooth in the escape wheel, the wheel gives the pallet a little push, which keeps the pendulum swinging.

The accuracy of a pendulum clock depends on two basic factors: How constant the length of the pendulum is, and how isolated the pendulum is. The first problem is a consideration only in the finest timepieces; very sophisticated pendulums have systems for keeping a constant length even if the pendulum expands or contracts in changing weather. The pendulum rod is made out of several materials, with differing rates of expansion. As the pendulum expands or contracts, the changes of length of the pieces of the pendulum rod cancel each other out, and the average length of the pendulum rod stays constant.

The second problem is really a question of "gilding the lily." How accurate does a clock have to be? The theory of a constant swing only applies to a free pendulum. During the time that the pendulum is getting kicked by the escape wheel it isn't running free. Hundreds of systems have been invented to make the moment of contact brief and sure. Many different geometries and ways of transmitting the energy have been used. In less perfect solutions, the pendulum is driven by the gear mechanism and the friction in the geartrain of the clock must be taken into account.

Pendulum clocks are only one type of clock. Before the pendulum was invented other mechanisms were used. Pendulum movements are not the answer for all applications. They will not work on board a ship because the housing of the clock must be stationary and a ship rolls with the tide, throwing the pendulum off count. There is no such thing as a pendulum wristwatch for the same reason. Other types of clock mechanisms have been invented for these purposes but they all share a basic design theme: Regular pulses are counted up and turned into a readable form.

The common digital clock counts up beats from a tiny vibrating piece of quartz crystal, and the most accurate clock in the world, a cesium atomic clock, counts up flashes of energy as radioactive material decays.

How a Clock Works

Knowing how the timer works and what each part does is important so that when you assemble the parts, you have a feel for how everything fits. A look at the movements of the pendulum and how the escapement train works in this model should provide an adequate explanation.

Assume that the escape wheel is trying at all times to move counterclockwise and the pendulum is in the middle of swinging right (fig. 3). This assumption is necessary because if the pendulum is at rest the clock won't move. The pendulum will keep swinging and the escape wheel will keep turning until a tooth of the escape wheel crashes into the left pallet which has swung into the path of the escape wheel (fig. 4). The timer stops dead while the pendulum finishes swinging high and starts to swing left. As the pendulum starts to move, the escape wheel pushes against the left pallet, trying to spin free. This extra push is what keeps the pendulum swinging. Finally the pendulum swings back to vertical and frees the escape wheel, which starts to turn again (fig. 5). At the top of the swing in the other direction, the right pallet catches an escape wheel tooth and again the clock movement is frozen (fig. 6). Then the pendulum finishes its swing and goes the other way, freeing the escape wheel to spin. As on the other side, the escape wheel gives the pallet another little push to keep it going and it pushes free. This cycle is repeated as long as the clock is wound.

The accuracy of the timer depends on how isolated the motion of the pendulum is from the force of the escape wheel, because when the escape wheel kicks the pallets, the force throws off the precise timing of the pendulum swings. In this model there is very little isolation, and when you finish building it you can sometimes see the pendulum being driven faster than it would normally go. I could have designed a more isolated movement but since the timer keeps fairly good time and it is easy to build, I consider it to be a successful design as is. In a sense, I ignored about three centuries of progress in return for something simple and reliable out of paper. There is a specific historical foundation for this design that I will discuss later, after the instructions for assembly.

One thing you may notice is that I have not mentioned the weights. This model has one weight, which hangs from the winding drum, which is part of the great wheel arbor. Its sole purpose is to overcome friction in the timer and turn the wheels; it plays no role in regulating the time other than producing the unwanted driving effect the escape wheel has on the pendulum.

The timer only has two gears: the great wheel, which is directly connected to the winding drum, and the escape wheel. A more complex clock could have several geartrains, each with many gears. One geartrain would move the hands and keep time, and other trains would set off any alarms or chimes. The great wheel transmits power to the escape wheel via a lantern pinion, which is a small diameter, wide-face gear.

The great wheel isn't *glued* directly to the winding drum, so that the timer can be rewound without interfering with the pendulum and escape wheel. There are two weight cords, wound in opposite directions. One cord holds the weights that power the clock. As this cord unwinds from the winding drum, the other cord, which has only a

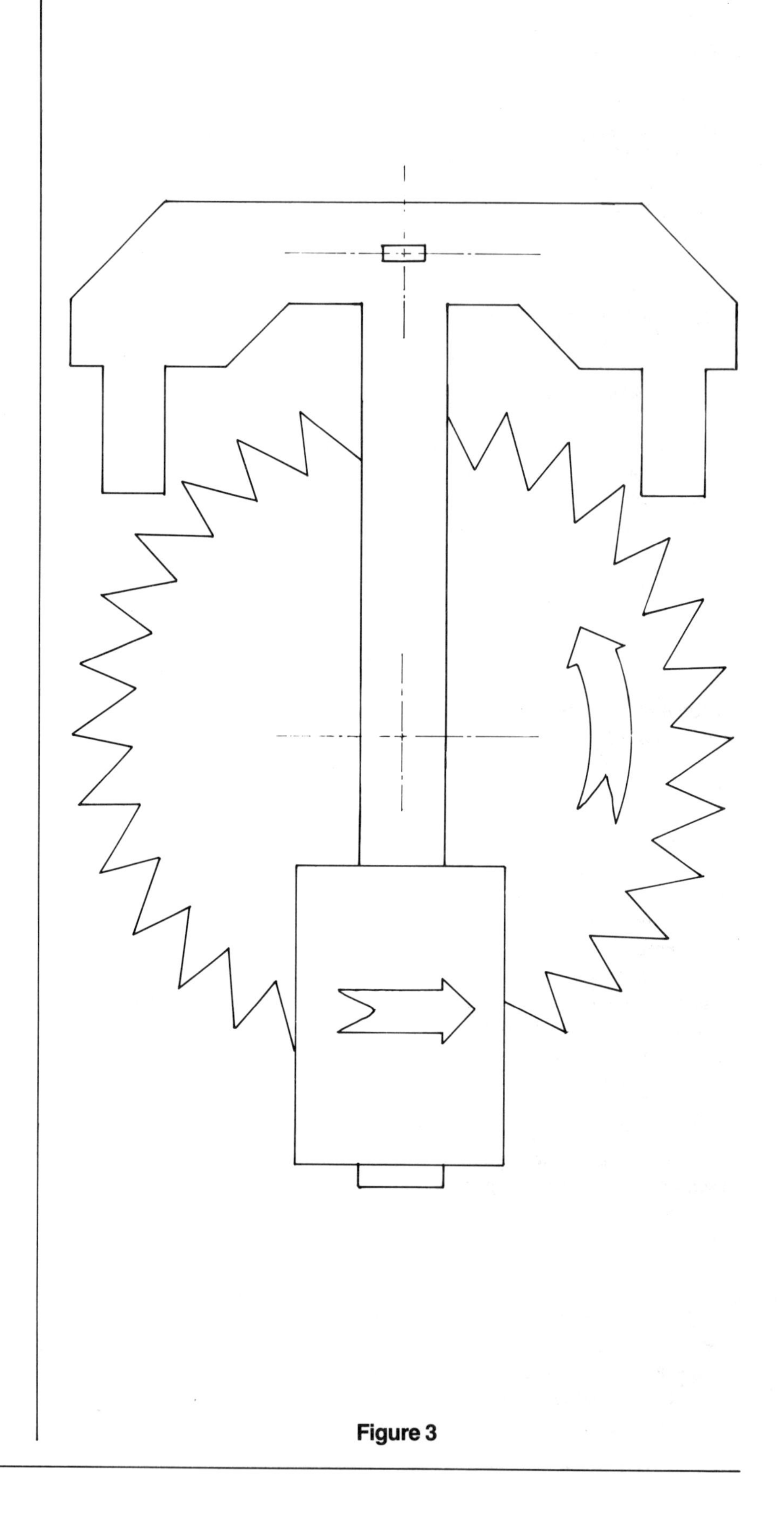

Figure 3

Figure 4

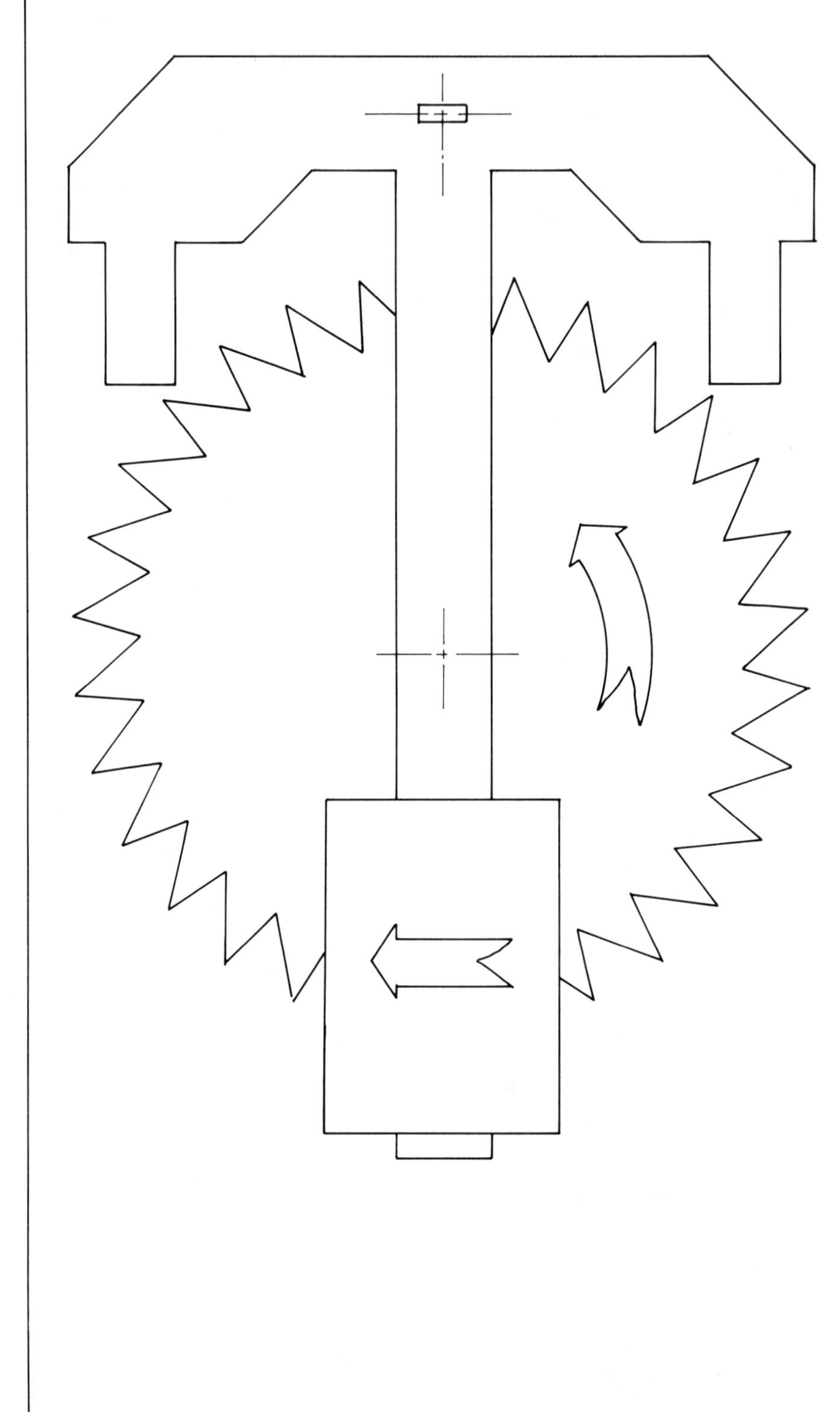

Figure 5

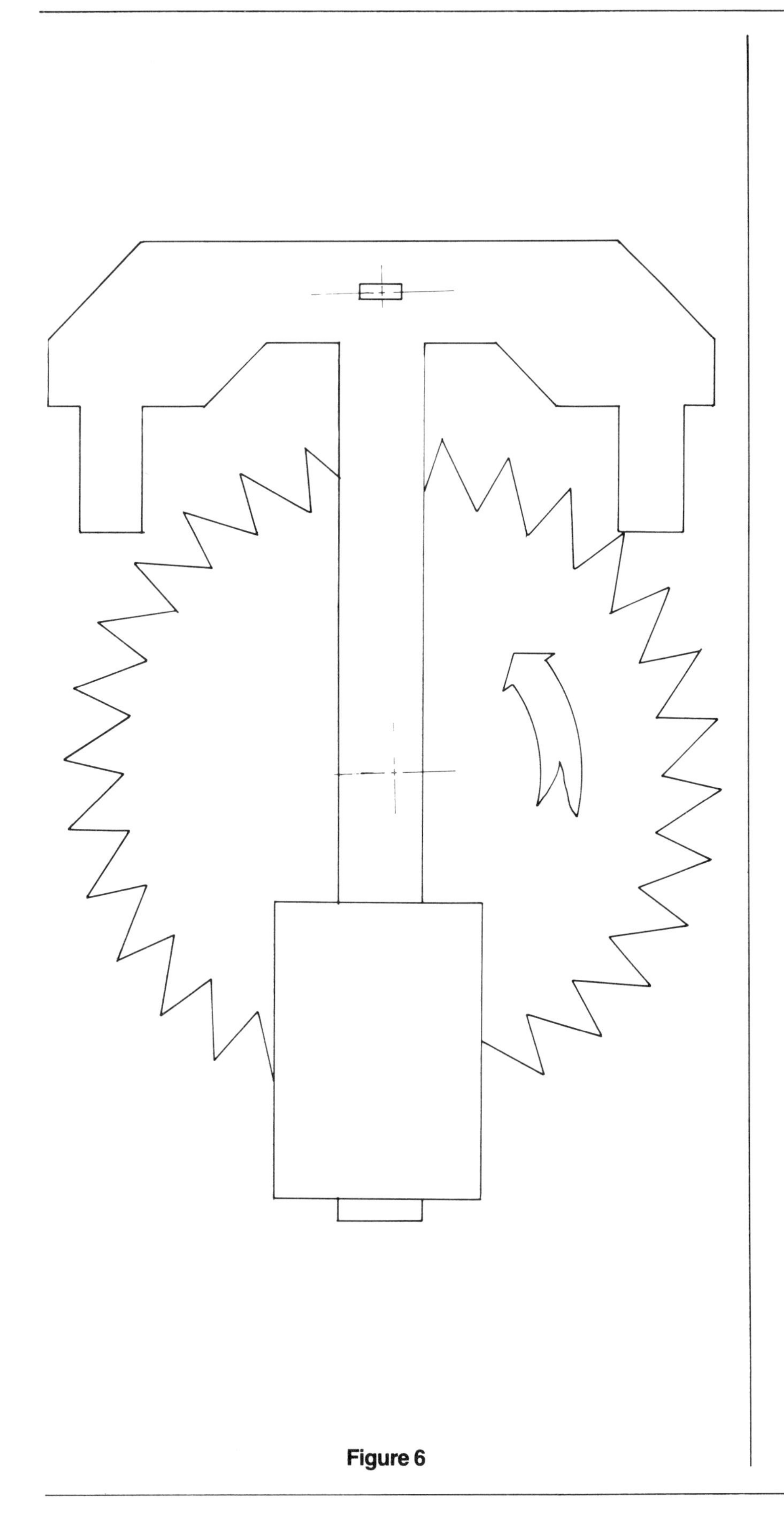

Figure 6

light weight on it to keep it straight, winds up. When the weight cord fully runs down, the other rewind cord is fully wound up, and pulling on it rewinds the main weighted cord. In order to be able to rewind the weights without turning the great wheel, which would change the timing and damage the escapement, the great wheel is connected to the winding drum by a ratchet. The ratchet makes it possible for the great wheel to turn with the winding drum in one direction only and to slip in the other direction.

The speed of the timer is determined by the length of the pendulum. The effective length of the pendulum can be changed by sliding the pendulum weight case up and down the pendulum rod. Sliding it up speeds up the timer and vice versa. The pendulum rod on this timer is long enough so that the great wheel can be adjusted to revolve exactly once every four minutes.

NOTES ON CONSTRUCTION

Tools and Equipment

There is absolutely no point in buying a kit for a few bucks and then having to shell out for a whole lot of tools. The only tools needed to assemble this model are a knife and a straightedge. The straightedge can be a ruler or a straight piece of metal that has a clean edge.

The paper parts must be detached from the back pages of the book and then folded. Use a sharp knife to cut away the little bridges that hold the parts to the pages. Cut them out slowly. Don't try to tear them free. If you accidentally slip with a knife and cut the paper in the wrong spot, just rub a little glue into the cut and let it dry. I usually cut out all the parts for a particular assembly at the same time, and I am very careful to store the cutout parts, the scrap, and the uncut pages in different areas so I don't accidentally use a good part for scrap or discard an uncut part.

I use a small X-Acto knife with disposable blades. The important thing in picking a knife is that the blade is sharp, which is why I like a hobby knife, where I can easily change blades. If you don't have a hobby knife you can use single-edged razor blades, which will work fine.

All the parts that have to be folded are prescored on the line where the bend is supposed to go. All the folds in the model are folded down. That is to say, with the printed surface of the part facing up, fold the bend away from you. On occasion you may find that the printed score line and the actual prescored crease are slightly out of alignment. This happens when the paper shifts while it is getting printed and precut at the factory. If you notice this condition, always cut or fold on the actual crease or precut line, **not the printed mark**.

The folds must all be clean and sharp, so do all your folding against a straightedge. Line up the score mark with the edge of the straightedge and prevent the part from shifting with your fingers, keeping your thumbs free. Then, pressing down firmly so that nothing moves, smoothly bend the fold over with your thumbs.

The only time this technique doesn't work is when you bend up the frames. The frames are too long to be bent in one shot, so work slowly from end to end and the bend will come out smoothly.

The parts are glued together with white glue. I used Elmer's Glue. I can't recommend any of the hobby glues or the instant glues because I have always had trouble with them. They dry too fast or they are hard to apply. White glue is easy to use, washes up with water and dries slowly to give you time to spread it around so you can get all the pieces correctly positioned. I never apply glue directly from the bottle. I use a spatula made from a scrap of paper so I can just get a touch of glue where I want it. After I apply the glue I squeeze the parts together with my fingers until the joint starts to dry. Sometimes, I clamp the parts together with a little piece of tape while they dry. After the glue dries I carefully peel off the tape so the paper underneath isn't damaged.

On the subject of tape: As a rule I always keep a roll of masking tape handy. I use it to clamp pieces together while glue is drying and hold parts down while I cut them up. The best kind to use is the masking tape that is used for drawings. You can peel it off when you're done without tearing the paper underneath. Never use cellophane tape because it's hard to remove.

You will also need pins for the arbors. The breakthrough that made this design feasible was figuring out how to make good arbors. In this model the arbors are made out of paper, with pins mounted in the center to give a low-friction bearing surface. The pins should be about one inch long. I used common dressmaker's straight pins that I bought at the local five and ten. If you prefer to raid a sewing kit, do so, but make sure that the pins you end up using are not bent. You need at least six pins. You will also need a bunch of pins that will be used to punch holes in the frames for the pivots. These pins can be bent. Get a few because they have a tendency to bend when you force them through the paper.

You will need about seven feet of string for the power string, rewind string, and to make a loop to hang the timer from a wall.

One of the things that bothered me was how to make and specify a standard weight for the weight case. I settled on pennies. You will need about fifteen pennies for the weight case, about three or four more for the pendulum, and one for the counterweight.

There are many illustrations scattered throughout the text. I tried to illustrate any operation that I thought anybody might have trouble with. Most of the drawings are in three-dimensional isometric projection, which is a technique that shows depth without going into exact mathematical perspective. Many times I didn't draw an entire piece and show only a small section "torn out" of the larger model. This was done to simplify the drawings in order to make them easier to understand.

Part Coding

All the parts are assigned a number. Identical parts share the same primary number with a suffix to show that the part is one of a set. Most gluing surfaces are marked. However, because the paper is printed on one side only, some glue areas are not shown. On some of the more critical parts where this occurs, printed on the side opposite the glue joint are notations such as "glue 45 here on other side." The text and illustrations should clear up any ambiguity about what goes where.

Figures 1 and 2 (see pages 8 and 9) show all the pieces and how they look assembled. Every piece in the model is shown in at least one illustration.

PUTTING THE PARTS TOGETHER

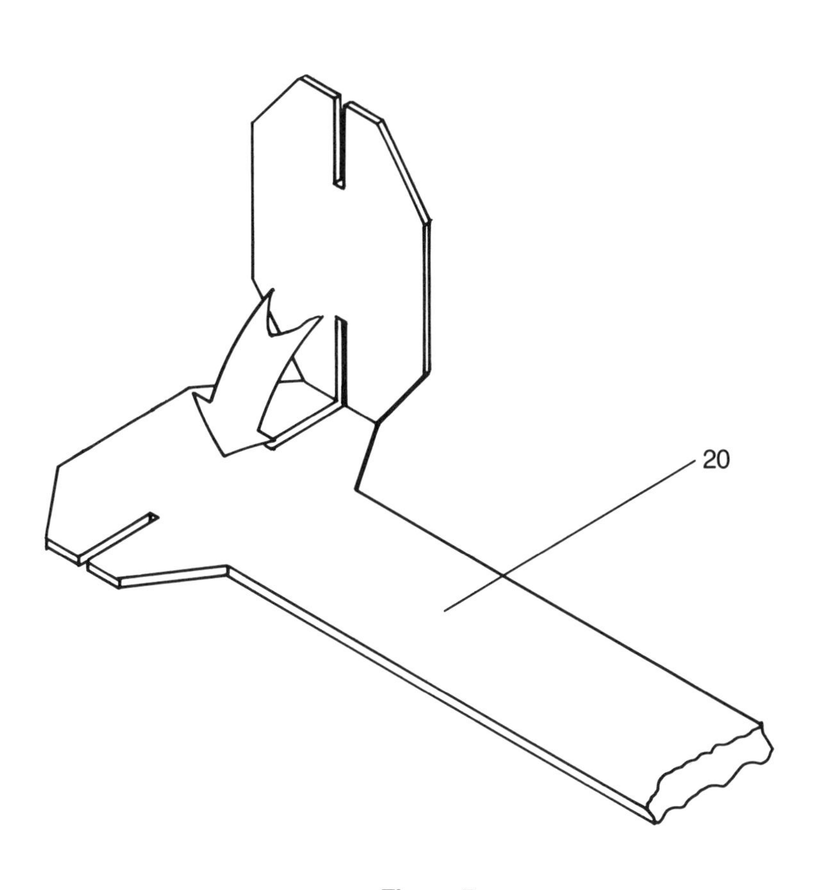

Figure 7

The Pendulum and the Weight Cases

My own personal rule of model-building is "always start with something simple." With that in mind, take the **Top Pendulum Rod [part 20]** and fold over and glue the upper support (fig. 7). Make sure that the slots line up and are free of glue.

The pendulum rod is composed of three segments, which are glued together. Glue the **Middle Pendulum Rod [part 21]** to part 20 and the **Bottom Pendulum Rod [part 22]** to the middle section (fig. 8). The parts are labeled to show how they are overlapped and glued. I found that the best way to ensure that the rod comes out straight is by holding it up on its edge and pressing the joints against a straight tabletop.

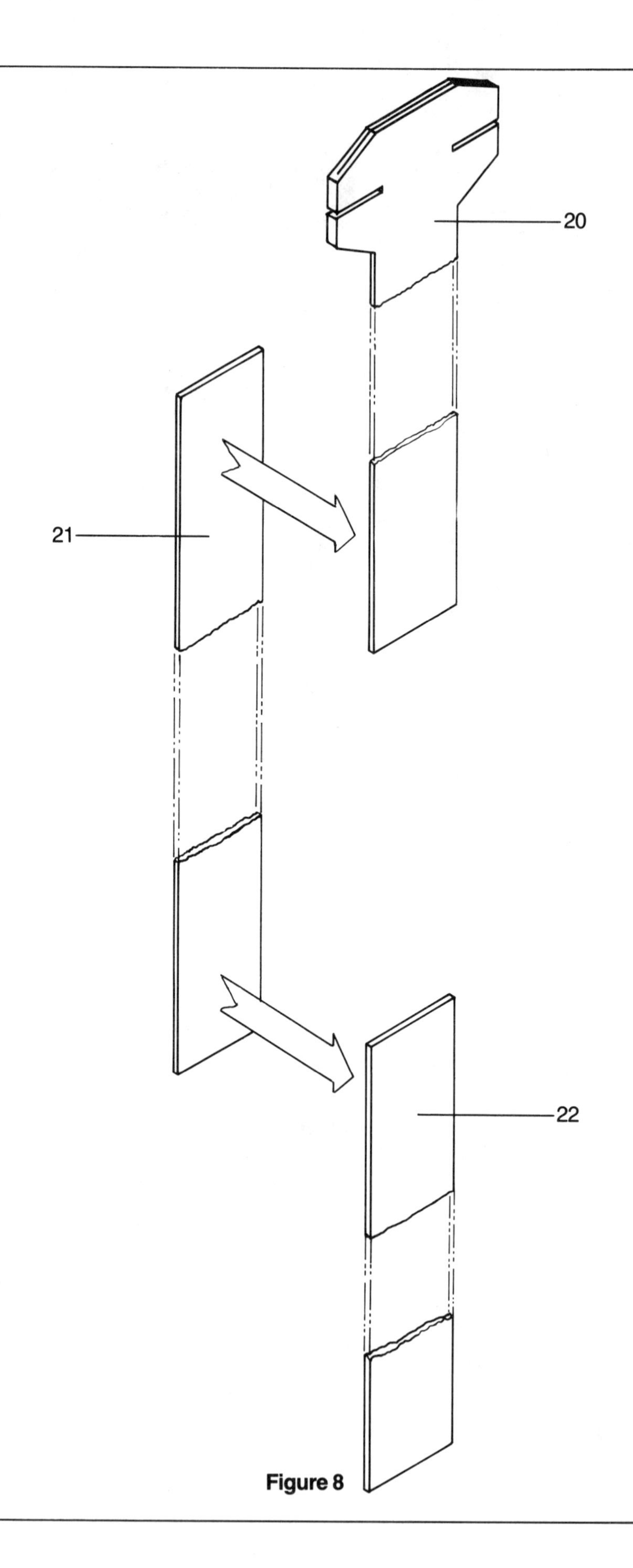

Figure 8

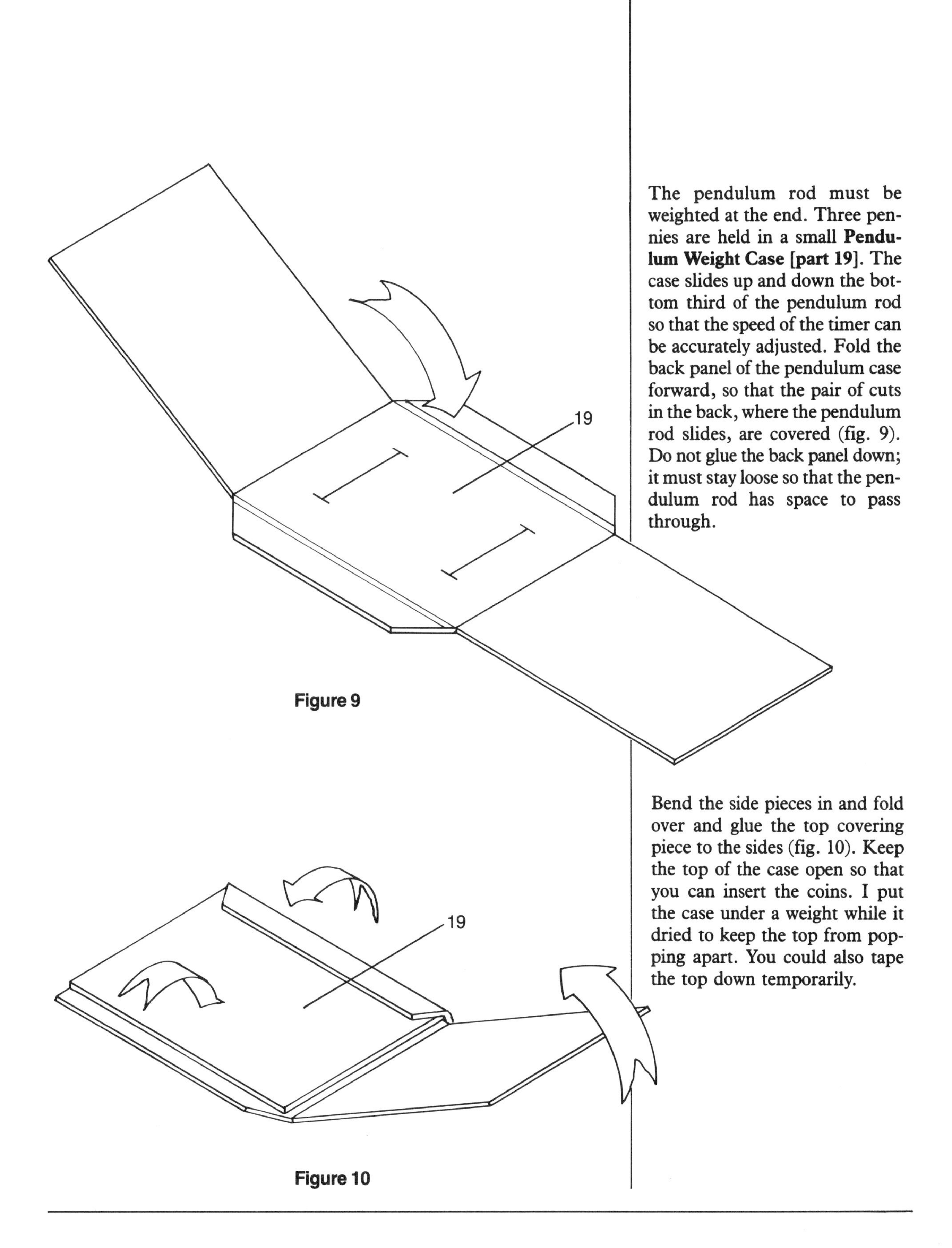

Figure 9

Figure 10

The pendulum rod must be weighted at the end. Three pennies are held in a small **Pendulum Weight Case [part 19]**. The case slides up and down the bottom third of the pendulum rod so that the speed of the timer can be accurately adjusted. Fold the back panel of the pendulum case forward, so that the pair of cuts in the back, where the pendulum rod slides, are covered (fig. 9). Do not glue the back panel down; it must stay loose so that the pendulum rod has space to pass through.

Bend the side pieces in and fold over and glue the top covering piece to the sides (fig. 10). Keep the top of the case open so that you can insert the coins. I put the case under a weight while it dried to keep the top from popping apart. You could also tape the top down temporarily.

Once the glue is dry, thread the case onto the pendulum rod (fig. 11) and then slide three pennies into the case (fig. 12).

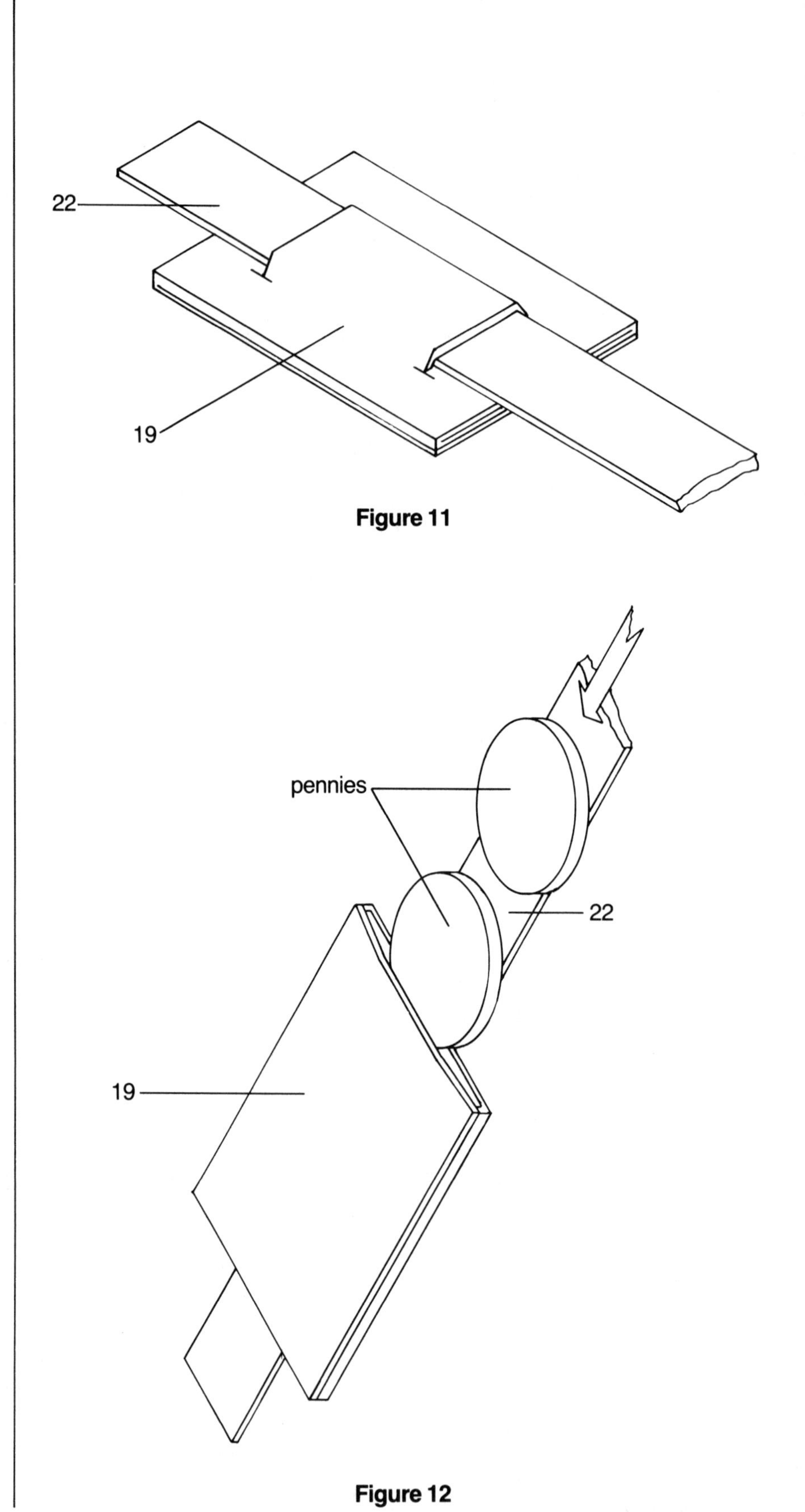

Figure 11

Figure 12

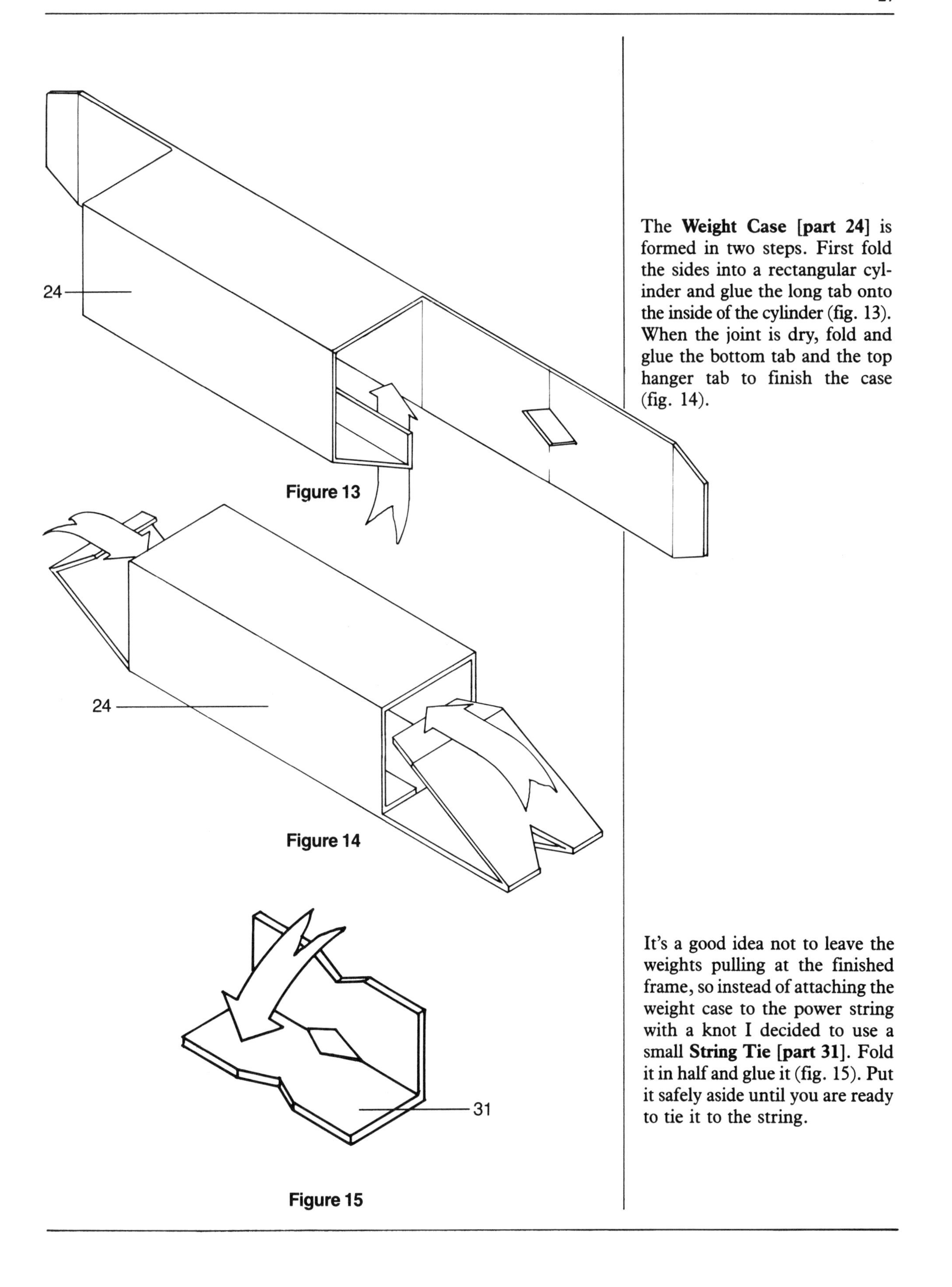

Figure 13

Figure 14

Figure 15

The **Weight Case [part 24]** is formed in two steps. First fold the sides into a rectangular cylinder and glue the long tab onto the inside of the cylinder (fig. 13). When the joint is dry, fold and glue the bottom tab and the top hanger tab to finish the case (fig. 14).

It's a good idea not to leave the weights pulling at the finished frame, so instead of attaching the weight case to the power string with a knot I decided to use a small **String Tie [part 31]**. Fold it in half and glue it (fig. 15). Put it safely aside until you are ready to tie it to the string.

The Great Wheel Arbor

Bend back both wings of both pieces of the **Great Wheel Arbor [parts 9–1 and 9–2]**, making sure that the center portion of each piece remains flat. I bent the wings against a steel ruler to ensure a very clean edge. Then glue the center portion of an arbor half to the center portion of the **Great Wheel Arbor Core [part 8]** (fig. 16). Do not put any glue on the bent wings. Make sure that the center core lines up with the outer arbor half in all directions. Drop a pin into each of the two slots that are cut into the arbor core. The pins form the actual pivots (fig. 17).

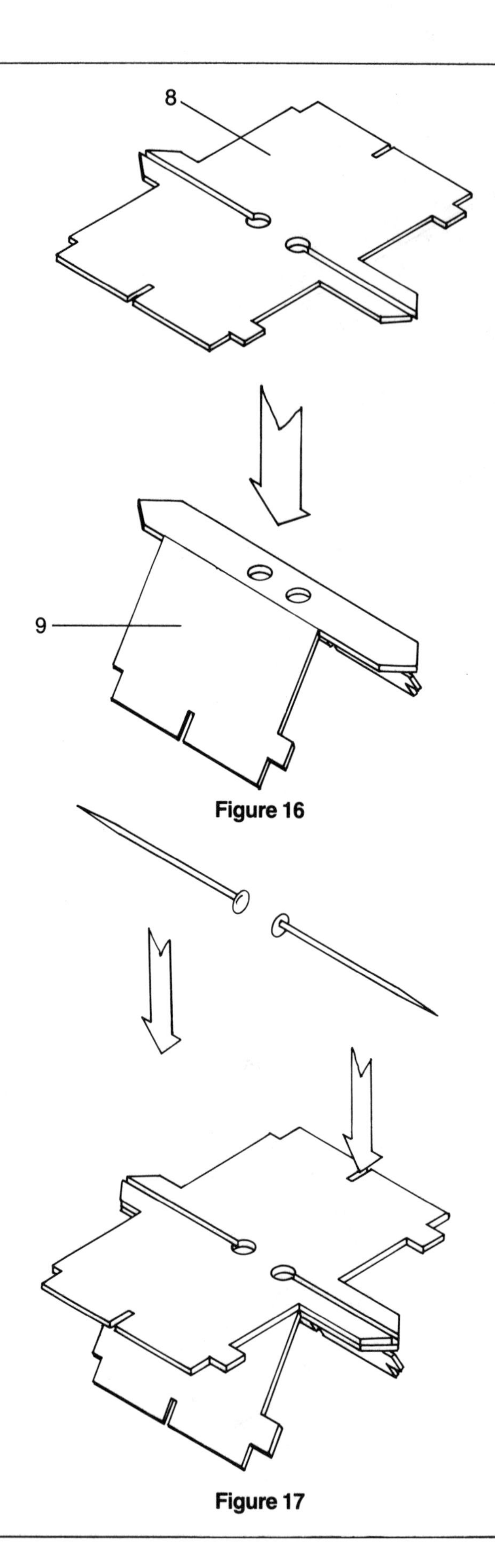

Figure 16

Figure 17

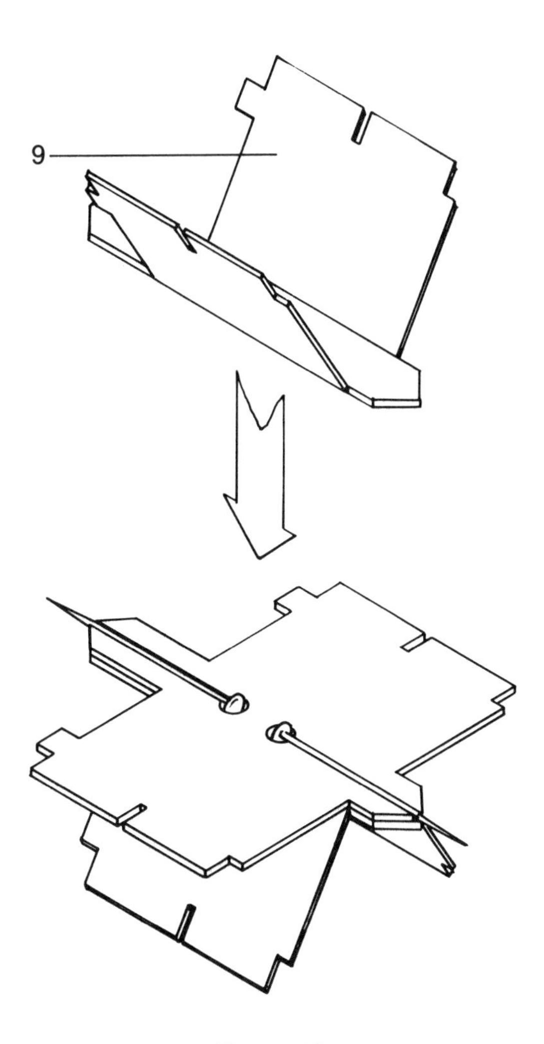

Figure 18

Add a little more glue to fill in the slots and then, taking care that the pieces are correctly aligned, glue on the other half of the arbor, sealing the pins in (fig. 18). Keep the parts from shifting by pressing the flat center section together with your fingers until the glue starts to set. Make sure that the wings are not stuck together when the glue dries. If any glue gets on the exposed portion of the pins, scrape it off with a knife after the arbor has dried. Cover the points of the pins with small pieces of masking tape, so that you don't get poked while you finish building the rest of the timer.

The wings of the arbor form a frame for the winding drum. The drum is separated into halves, one for the power string and one for the rewind string. The **Winding Drum Spacer [parts 27–1 and 27–2]** is installed in two sections. Take each section, line it up, and glue it into the notches cut out in the center of the arbor wings (fig. 19). Glue or tape the edges of the halves together to form a smooth ring so that the winding strings will not catch on the joints. The **Winding Drum Rings [parts 6–1 and 6–2]** are glued to each end of the drum to keep the winding strings from falling off (fig. 20).

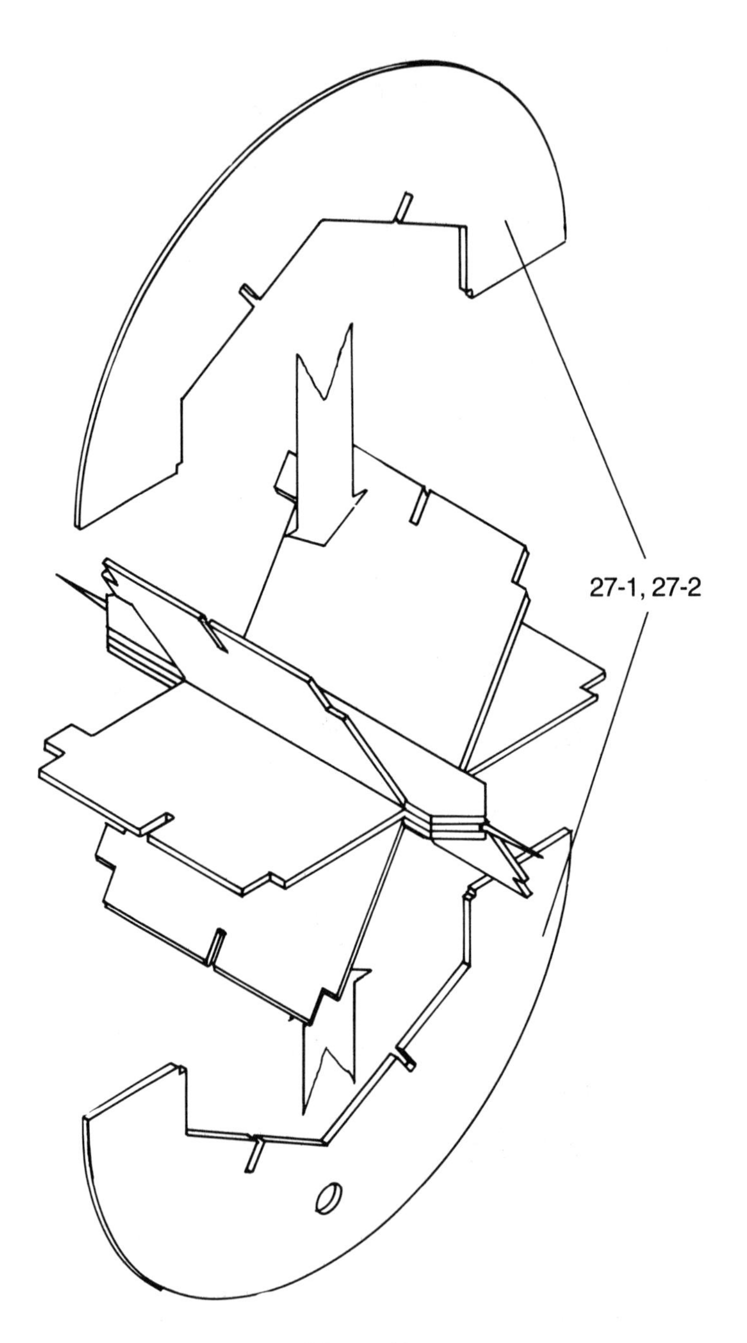

Figure 19

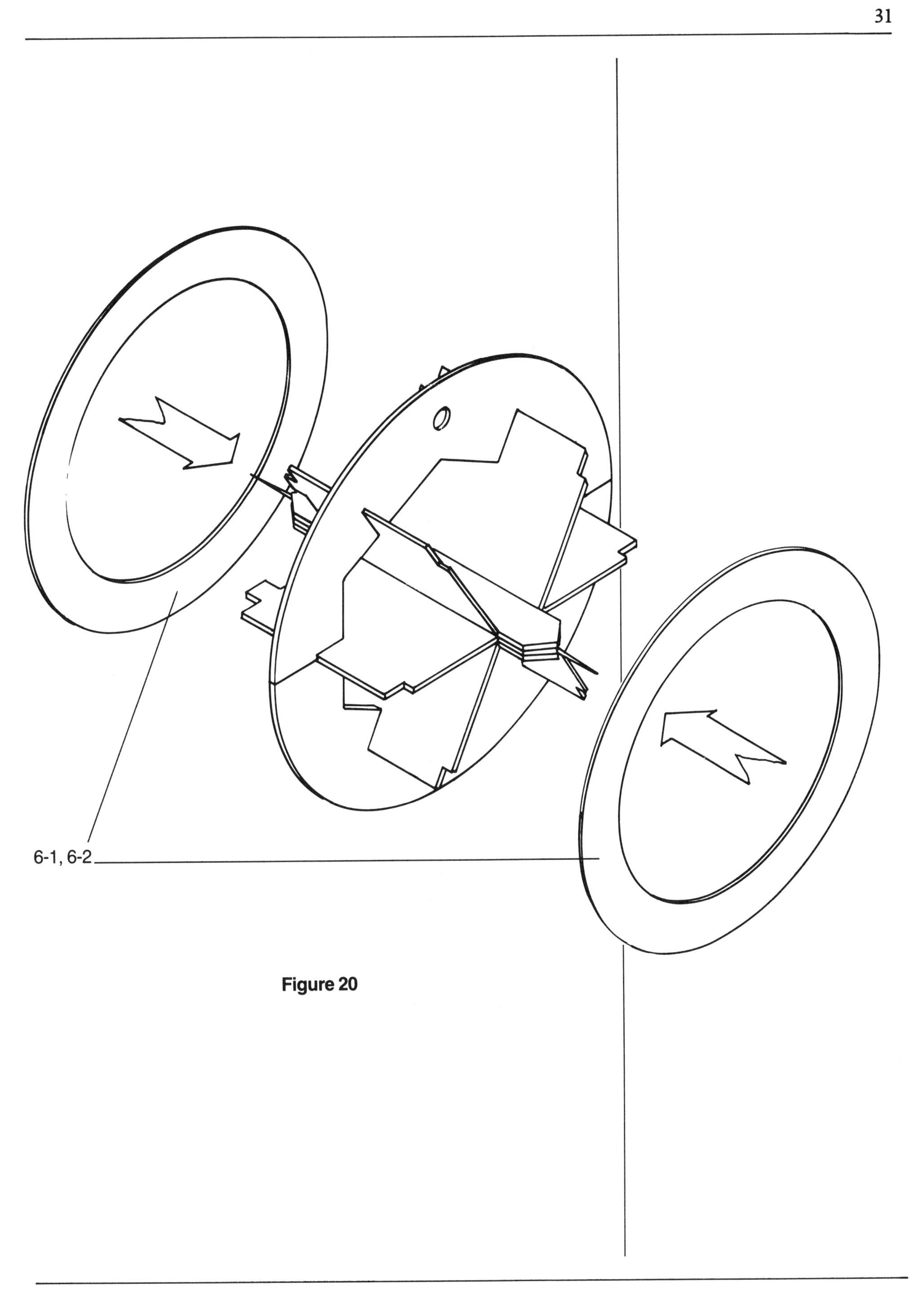

Figure 20

This is a convenient time to install the winding strings. Take a 6-foot length of smooth string and fold it in half to find the center point. (You can use more than six feet; the length is governed by how much room the weight case has to fall in.) Thread the string through the small hole that is near the rim of one of the center spacers. Center the string, with equal lengths on each side of the spacer, and tape it or glue it in place so that the string on each side of the hole points toward the drum. Wind the string that falls on the rewind side of the spacer around the drum in a counterclockwise direction. There are notations printed on the drum wings that show which side is which. The rewind drum is at the front of the arbor and you should wind in a counterclockwise direction as you look at the front of the arbor. At the free end of the rewind string tape a penny so that when the timer is running the string will hang neatly down. Tie the other string, which is the power string, to the string tie. Don't wind it around the other drum but, for now, you may want to tape both strings out of the way so they don't get tangled.

Carefully, without damaging any of the teeth, cut out the **Great Wheel [part 25]**. This wheel will ride, free to turn, on the great wheel arbor, so do not glue it in place. The winding drum transmits power to the great wheel through a ratchet system so the wheel will turn under power in one direction and rewind freely, without binding the rest of the geartrain, in the other.

The **Ratchet Wheel [parts 7–1, 7–2, 7–3, and 7–4]** is made up of four identical pieces that are glued together to form a single thick sandwich. When you glue the layers together make sure they all have the ratchet going in the same direction and that the rectangular arbor holes in the center of the pieces are uniformly aligned (fig. 21). I aligned the ratchet from the inside. Before the glue set, I took the tip of my knife, held it flat, and pressed against the edges of all of the leaves from the inside of the arbor hole. Test fit the ratchet on the great wheel arbor, just to make sure it fits, but *do not glue it in place*, and remove it from the arbor before the glue dries and it gets stuck on.

While the ratchet wheel dries, fold and glue the **Ratchet Clicks [parts 12–1 and 12–2]** in half (fig. 22).

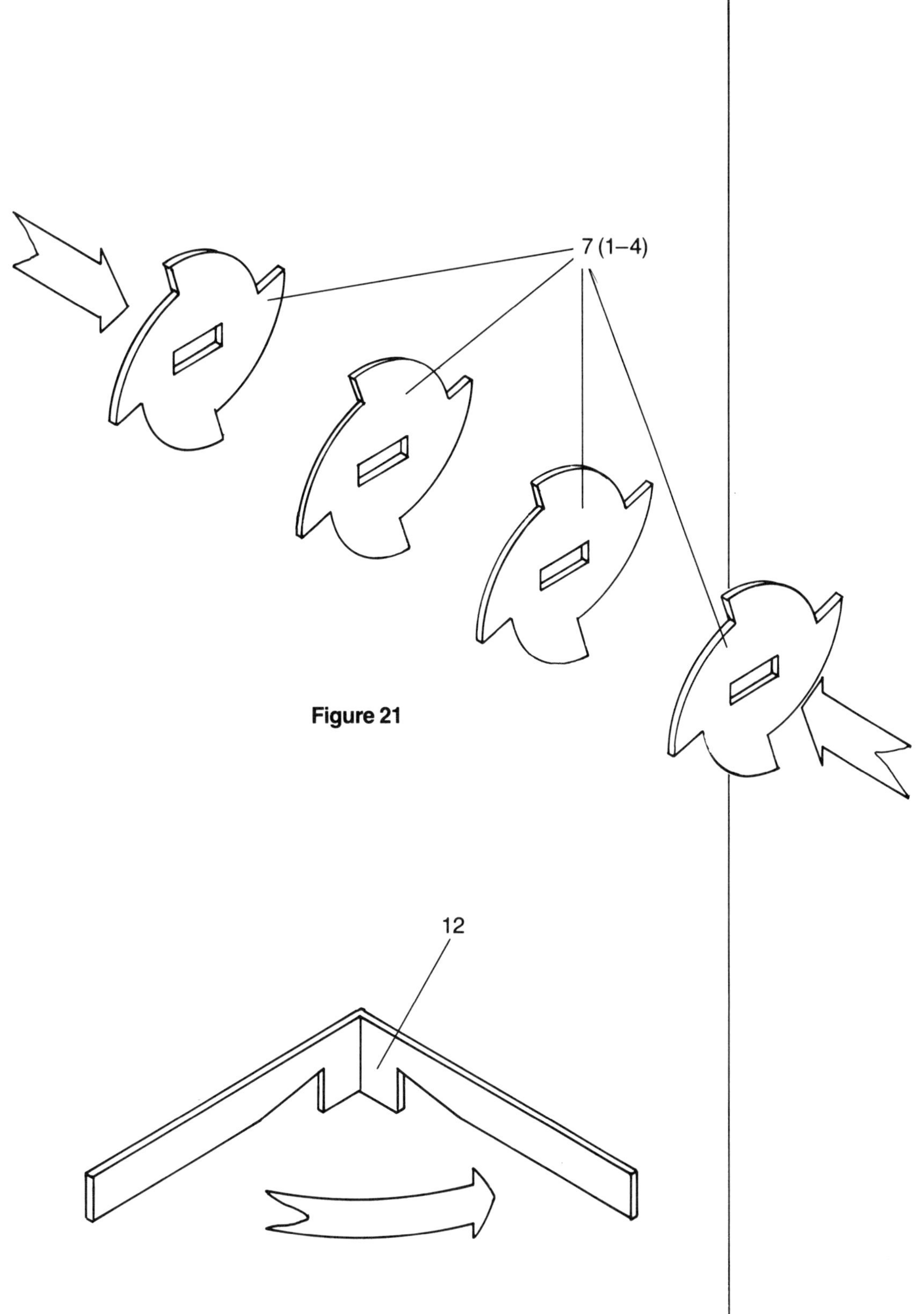

Figure 21

Figure 22

Glue one click into the cutout in the great wheel that is directly under the number 2 (fig. 23). Glue the click into the side of the wheel **without** any printing. Only apply glue to the cutout area, because the body of the click must be free to bend.

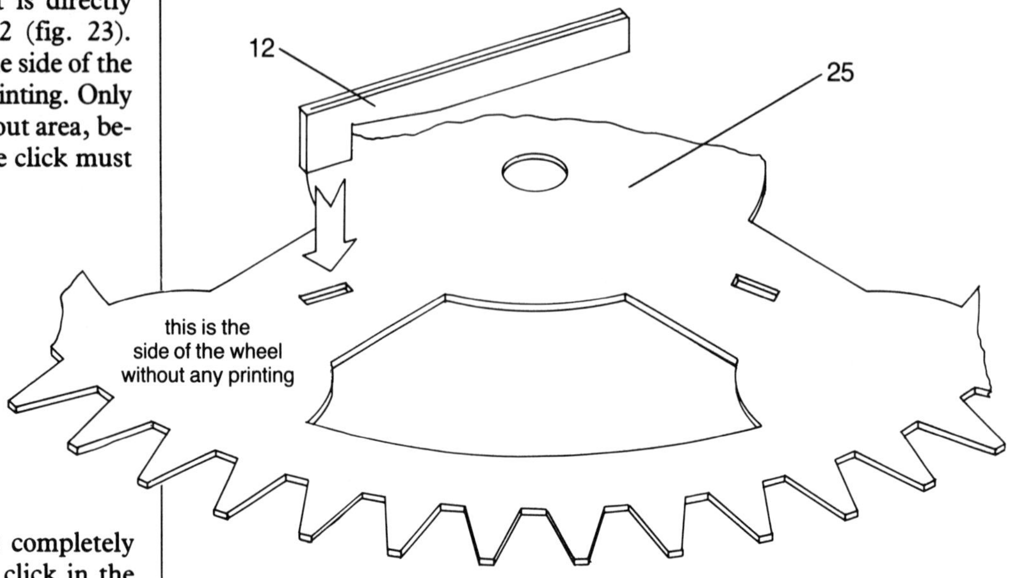

12
25
this is the
side of the wheel
without any printing

Figure 23

After the first click completely dries glue the other click in the second slot under the number 1 (fig. 24). Again the click is glued into the unprinted side of the wheel. It is important to let the first click completely dry in its slot, so that when you install the second click it will be correctly pushed into tension by the first.

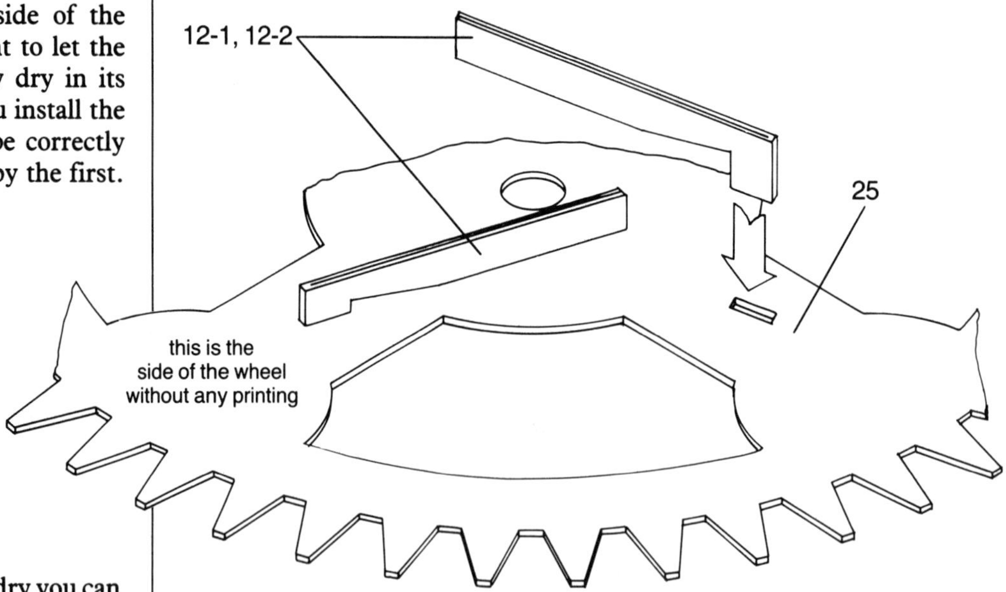

12-1, 12-2
25
this is the
side of the wheel
without any printing

Figure 24

When both clicks are dry you can complete the great wheel arbor assembly by sliding the great wheel onto the rear (longer) part of the arbor, so that the printing faces the winding drum. The great wheel spins on the arbor, held away from the back drum ring by the little nibs sticking out from under the ring. Slide the ratchet onto the arbor, in back of the wheel, so that the great wheel cannot slide off. Gently bend the clicks away and slide the ratchet under them so that the ratchet is engaged. Make sure you install the ratchet with the orientation shown in figure 25.

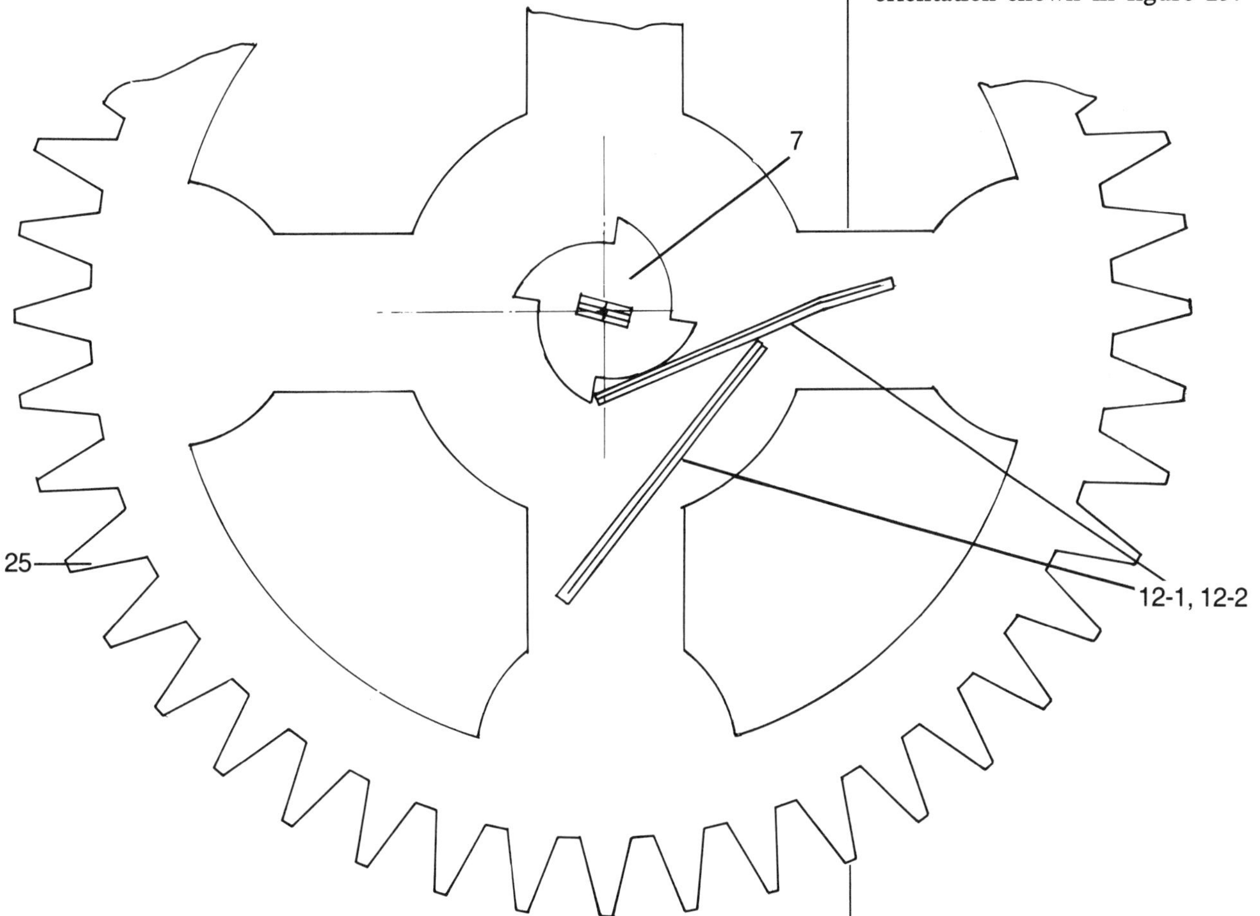

Figure 25

You do not need to glue the ratchet down. If you do decide to glue it to the arbor make sure that the great wheel is still free to turn. Notice that the winding drum now will only turn free in one direction; in the other direction, the ratchet on the arbor catches the great wheel, and both parts turn as one.

The Escape Arbor

The escape wheel assembly starts as a three-layer arbor sandwich like the great wheel. Take the **Escape Wheel Arbor Core [part 1]** and glue it to one of the two **Escape Wheel Arbors [parts 3-1 and 3-2]**. Bend down the wings on the arbor using a straightedge to get a clean bend, as you did for the great wheel. Glue one arbor to the arbor core, aligning the two parts so that the pointed ends line up and the raised shoulders match (fig. 26).

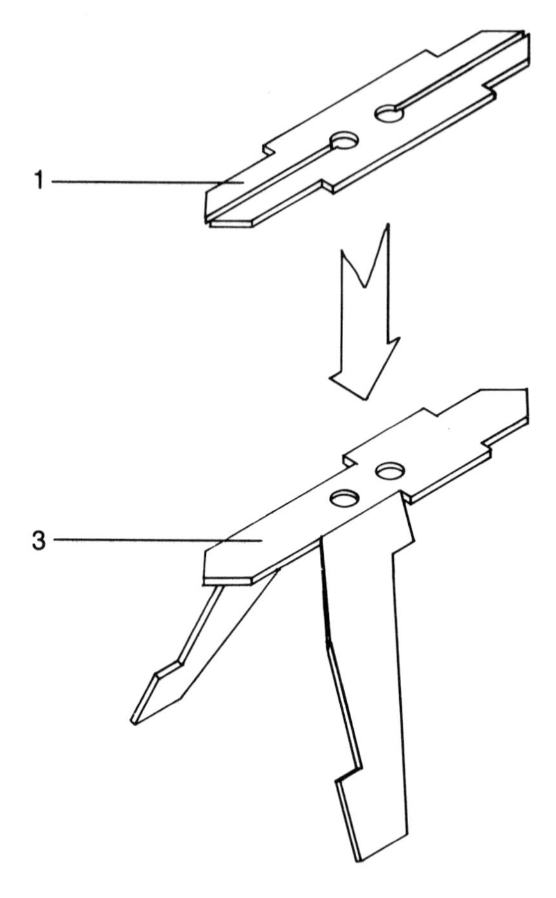

Figure 26

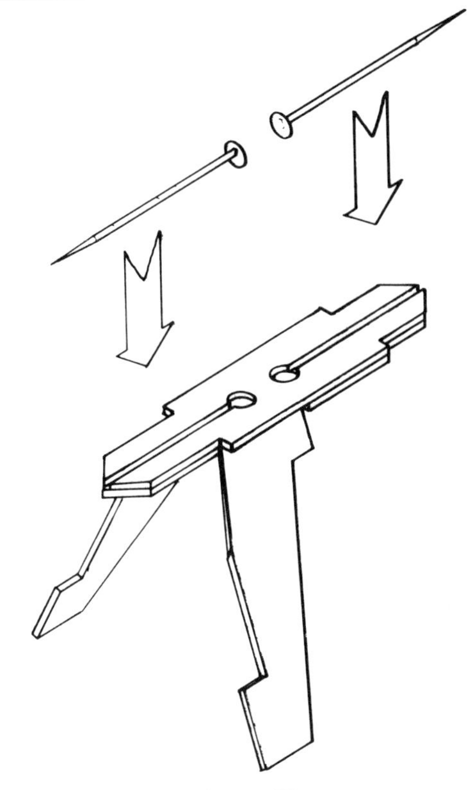

Figure 27

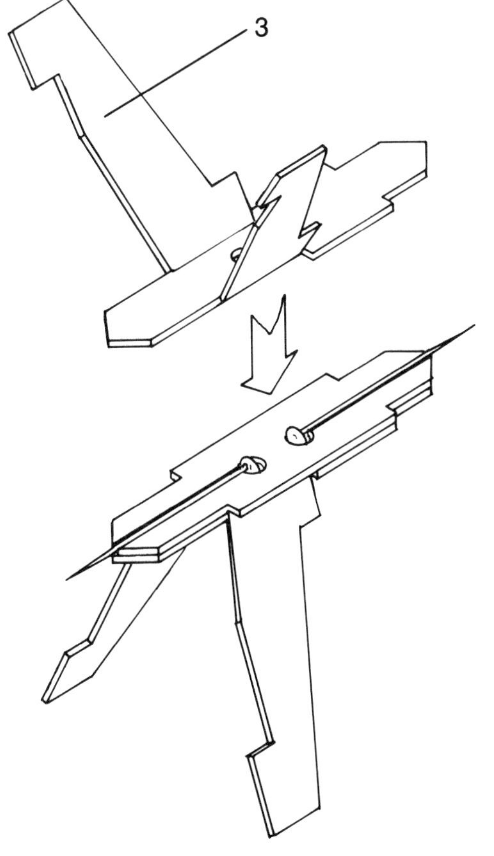

Figure 28

Drop a pin into each of the slots formed by the core (fig. 27) and then seal the pins in by gluing the second arbor piece over the core (fig. 28).

Carefully, so the teeth are not damaged, slide the **Escape Wheel [part 26]** onto the end of the arbor. Spread the arbor wings so that they key into the slots in the wheel (fig. 29). **Be sure that the escape wheel is correctly installed with the unprinted side glued against the supports and the printed side facing forward.** The wheel must be seated against all four supports.

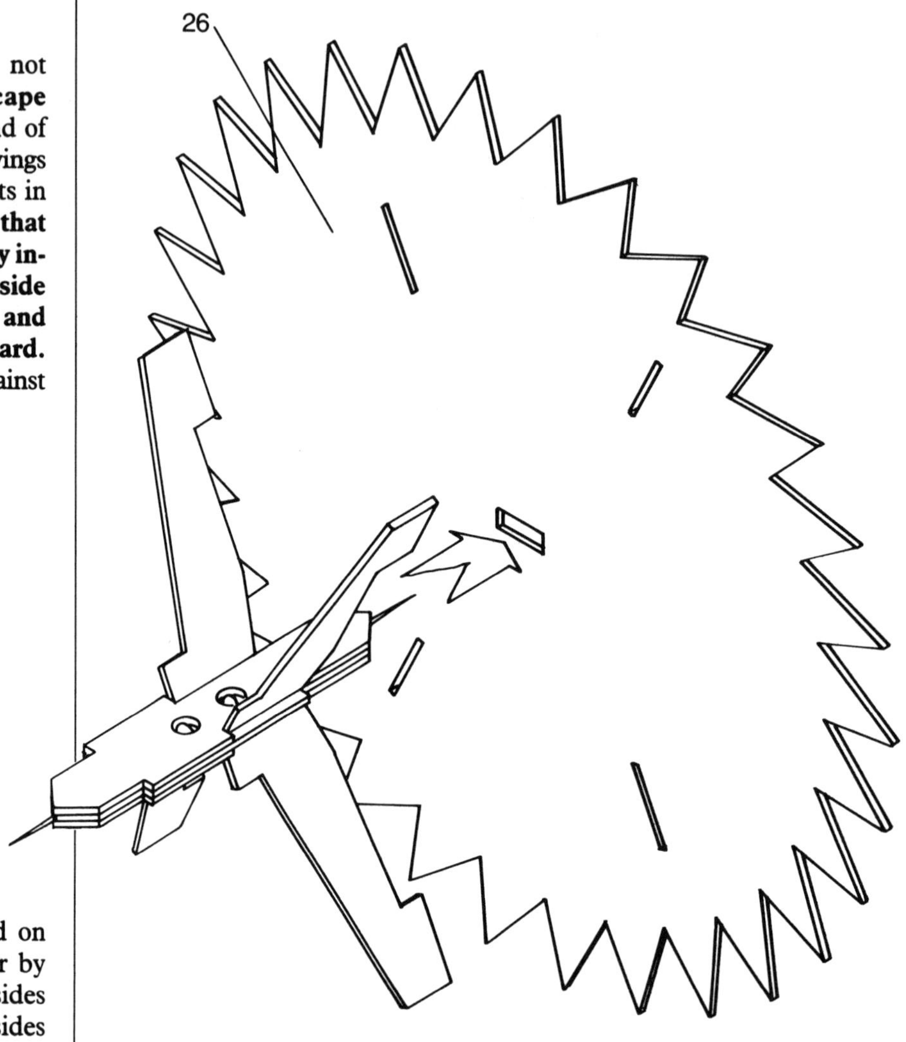

The pinion wheel is formed on the back of the escape arbor by first sliding the two pinion sides on. After you cut out the sides from the body of the book, clean out the slots for the gear teeth so that all the teeth seat equally.

26

Figure 29

Slide the **Inner Pinion Side [part 5]** on first. It seats against the far step on the arbor. The teeth must be angled in the correct direction, so make sure the printed surface faces the escape wheel and the unprinted surface faces the back. Slide the **Outer Pinion Side [part 4]** so that it seats against the first step in the arbor.

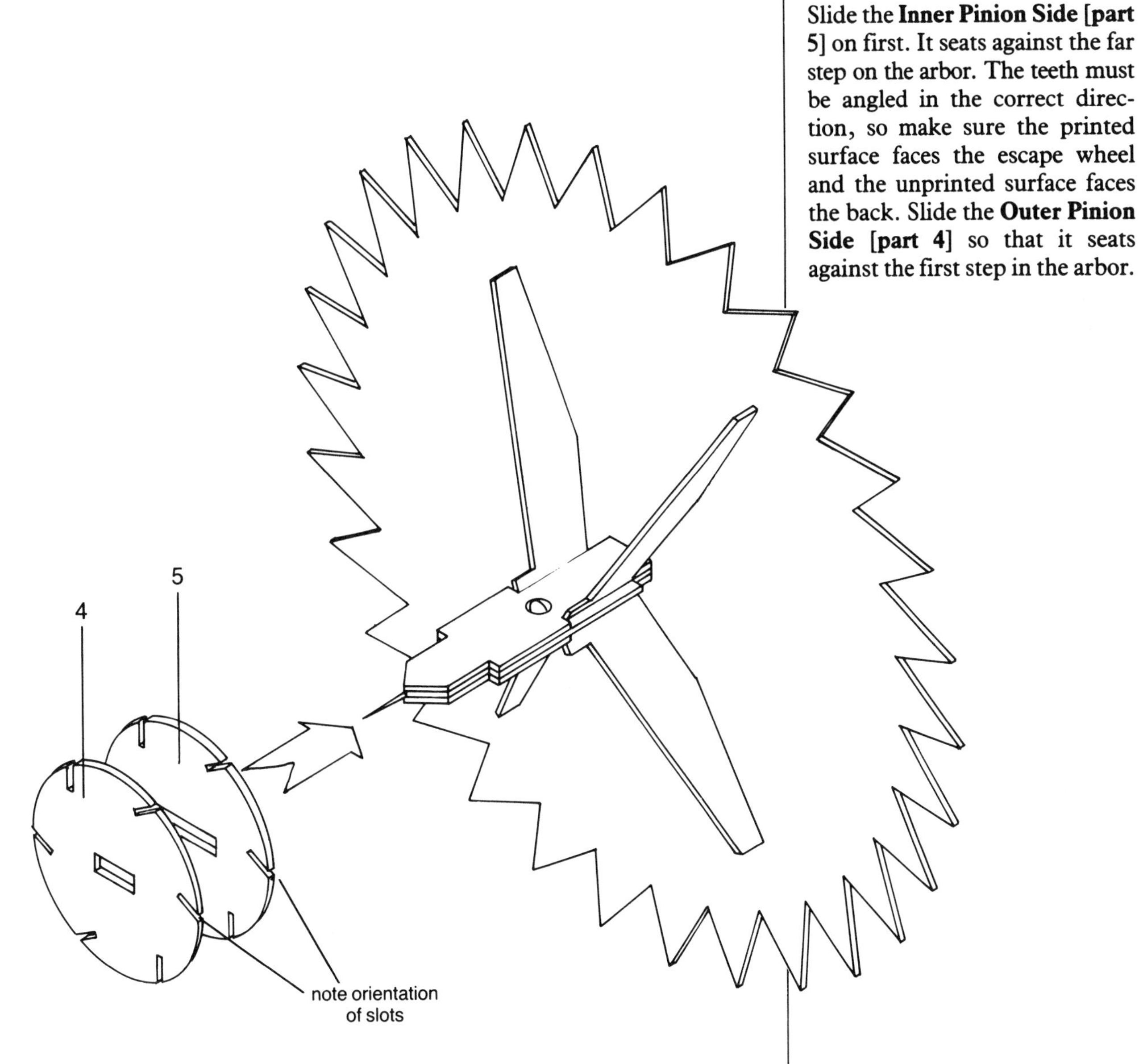

Figure 30

Again, make sure that it is correctly oriented and that the printed surface faces the escape wheel. The slots for the pinion wheel teeth should line up (fig. 30).

Slide the **Pinion Gear Teeth [parts 2–1 to 2–6]** into the six slots in the pinion sides (fig. 31). If you do this before the glue on the pinion sides completely dries, any slight tilt in the sides will be corrected. Be certain that the teeth are firmly seated in their slots so that the gear is even and later will run smoothly. An extra pinion tooth is included in the kit in case one tooth gets damaged during assembly.

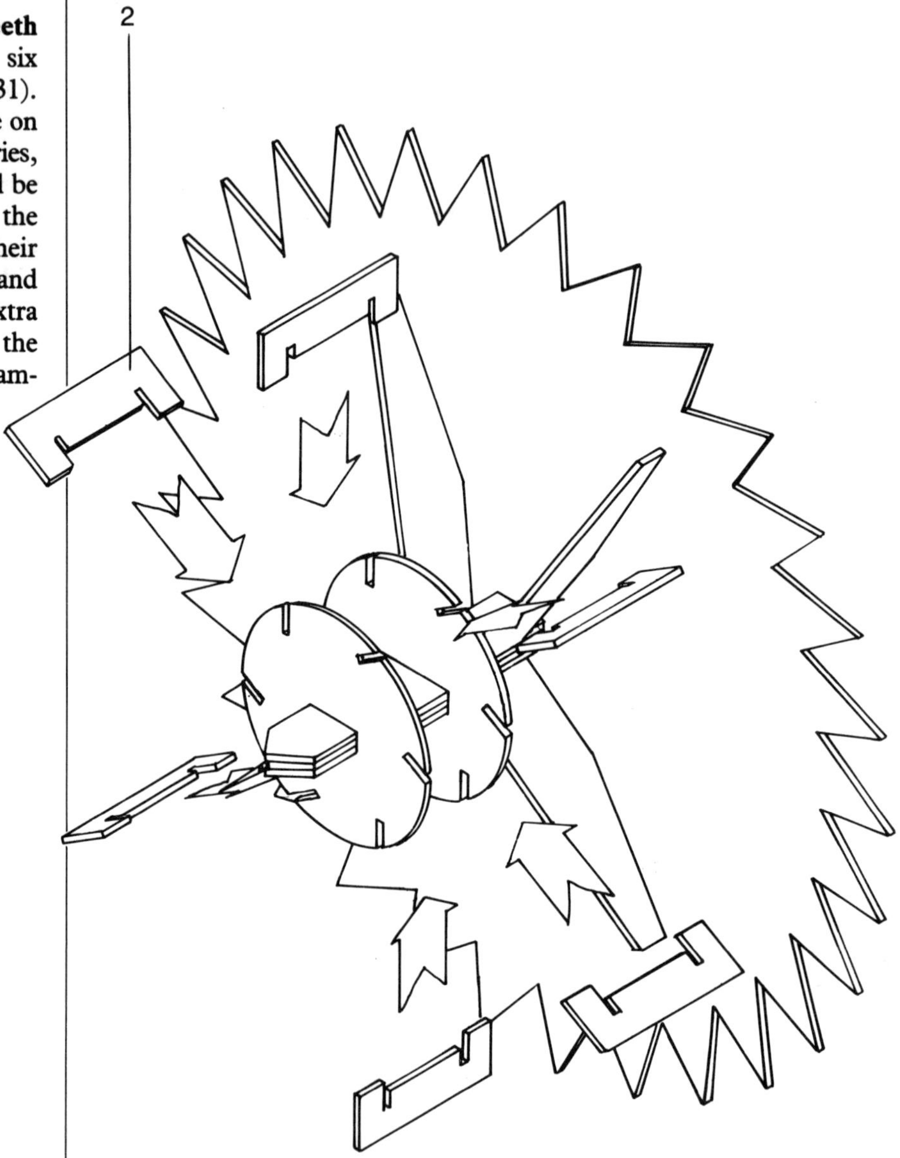

Figure 31

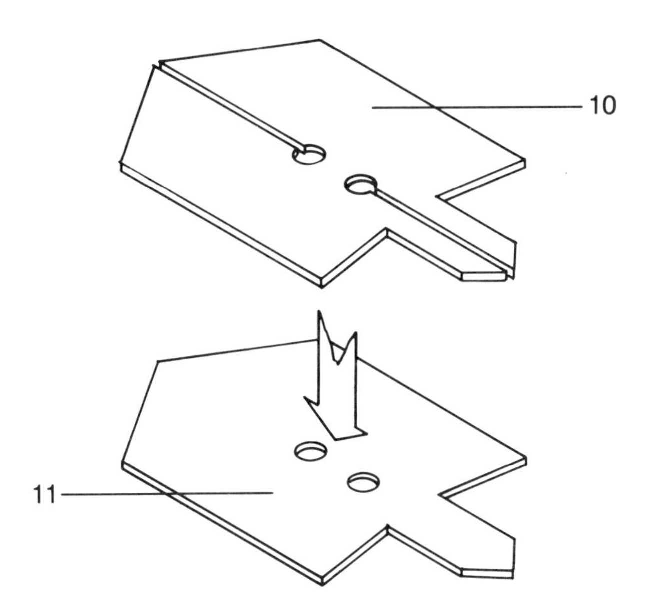

Figure 32

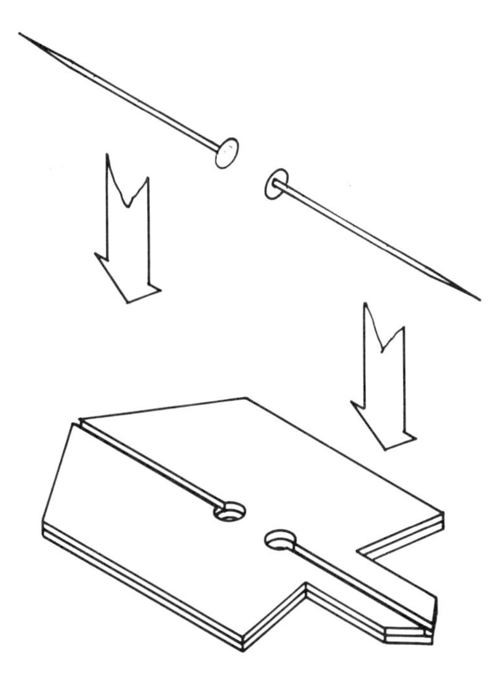

Figure 33

The Pallet Arbor

The pallet arbor is also composed of three layers, and your first job is to glue the **Pallet Arbor Core [part 10]** to the **Pallet Arbor Bottom [part 11]** (fig. 32). Then drop a pin into each channel in the arbor core to form the pivots (fig. 33).

Bend the sides of the **Pallet Arbor Top [part 28]** down (fig. 34) and then glue the arbor core and bottom to the underside of the arbor top, sealing the pins in (fig. 35). Keep the three parts aligned so that they form a smooth, solid core.

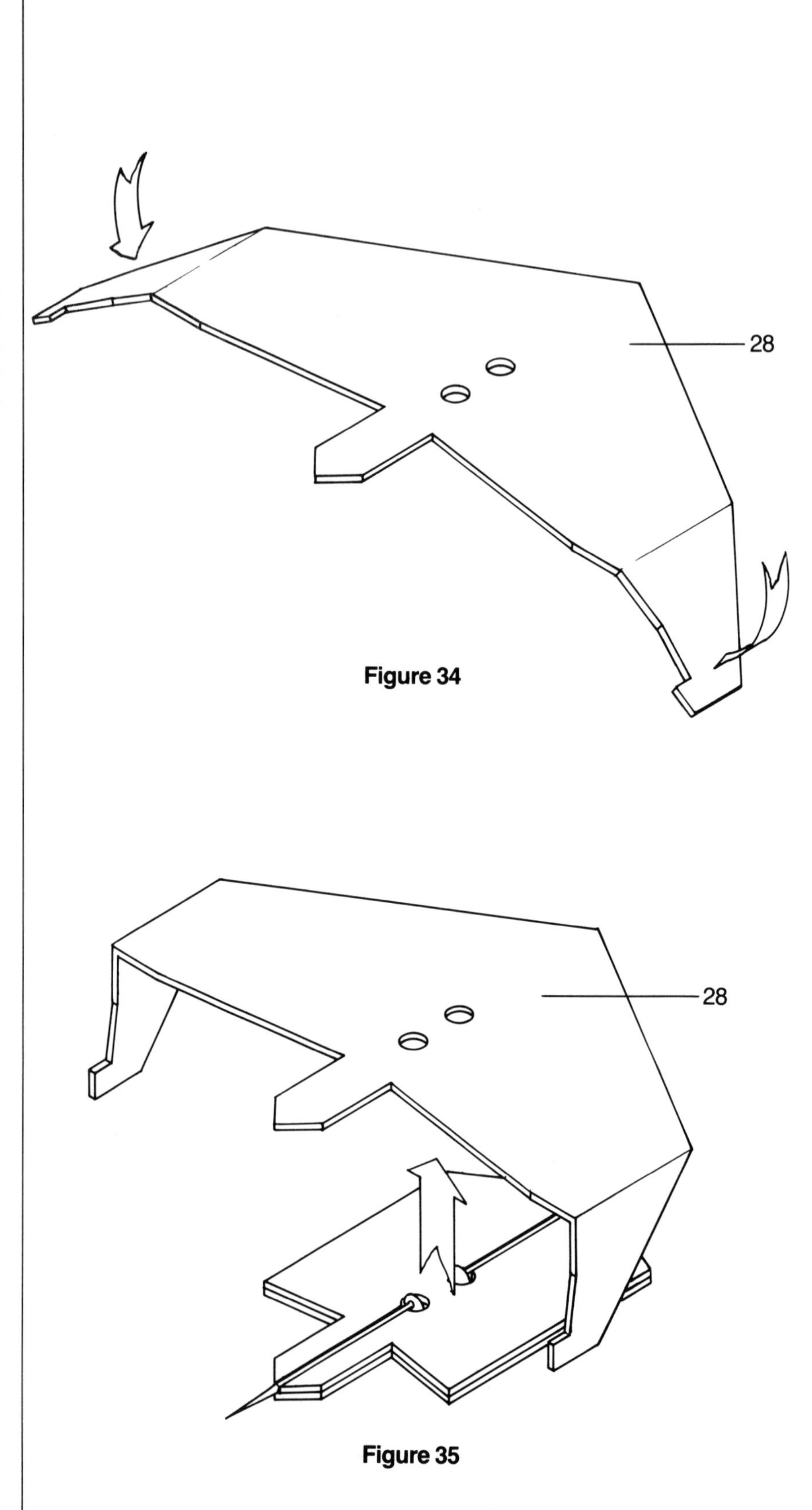

Figure 34

Figure 35

The pallet arbor supports the **Pallet Arm [part 17]** but before you glue the arm on, you must first attach both **Pallet Teeth [parts 18–1 and 18–2]** to the arm. Each pallet tooth has two folds which should be prefolded so that they turn out nice and sharp. Install each tooth by first sliding the side with the center slot over and into the outside notch on the pallet arm. The tooth interlocks with the vertical notch and fits around the small key on the arm. Then slide the free end of the tooth so that it interlocks with the notch on the inside of the arm (fig. 36). Make sure that the teeth are firmly seated against the arm. I temporarily clamped them to the arm with a small piece of tape to keep the teeth from slipping out while the glue was drying.

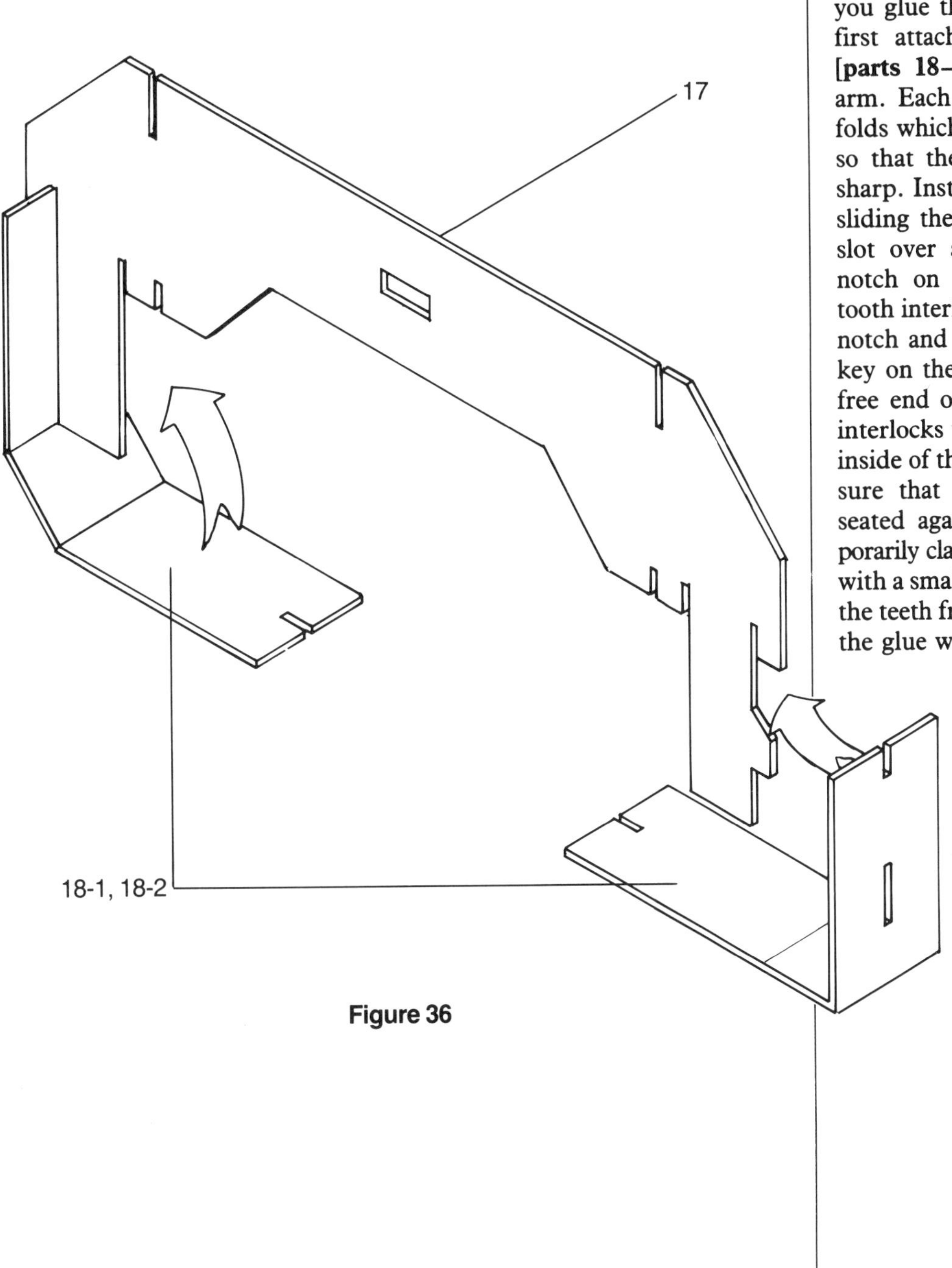

Figure 36

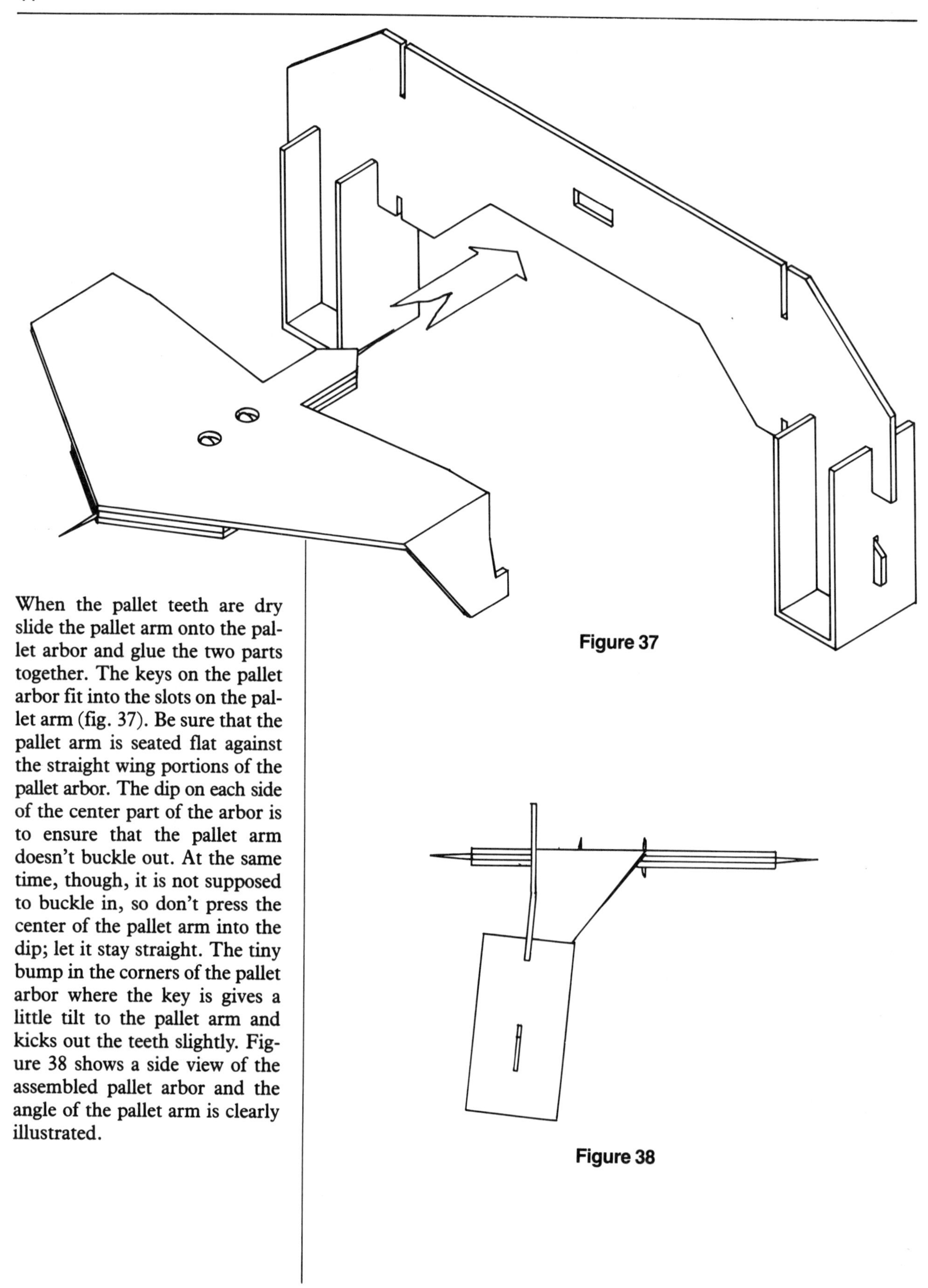

When the pallet teeth are dry slide the pallet arm onto the pallet arbor and glue the two parts together. The keys on the pallet arbor fit into the slots on the pallet arm (fig. 37). Be sure that the pallet arm is seated flat against the straight wing portions of the pallet arbor. The dip on each side of the center part of the arbor is to ensure that the pallet arm doesn't buckle out. At the same time, though, it is not supposed to buckle in, so don't press the center of the pallet arm into the dip; let it stay straight. The tiny bump in the corners of the pallet arbor where the key is gives a little tilt to the pallet arm and kicks out the teeth slightly. Figure 38 shows a side view of the assembled pallet arbor and the angle of the pallet arm is clearly illustrated.

Figure 37

Figure 38

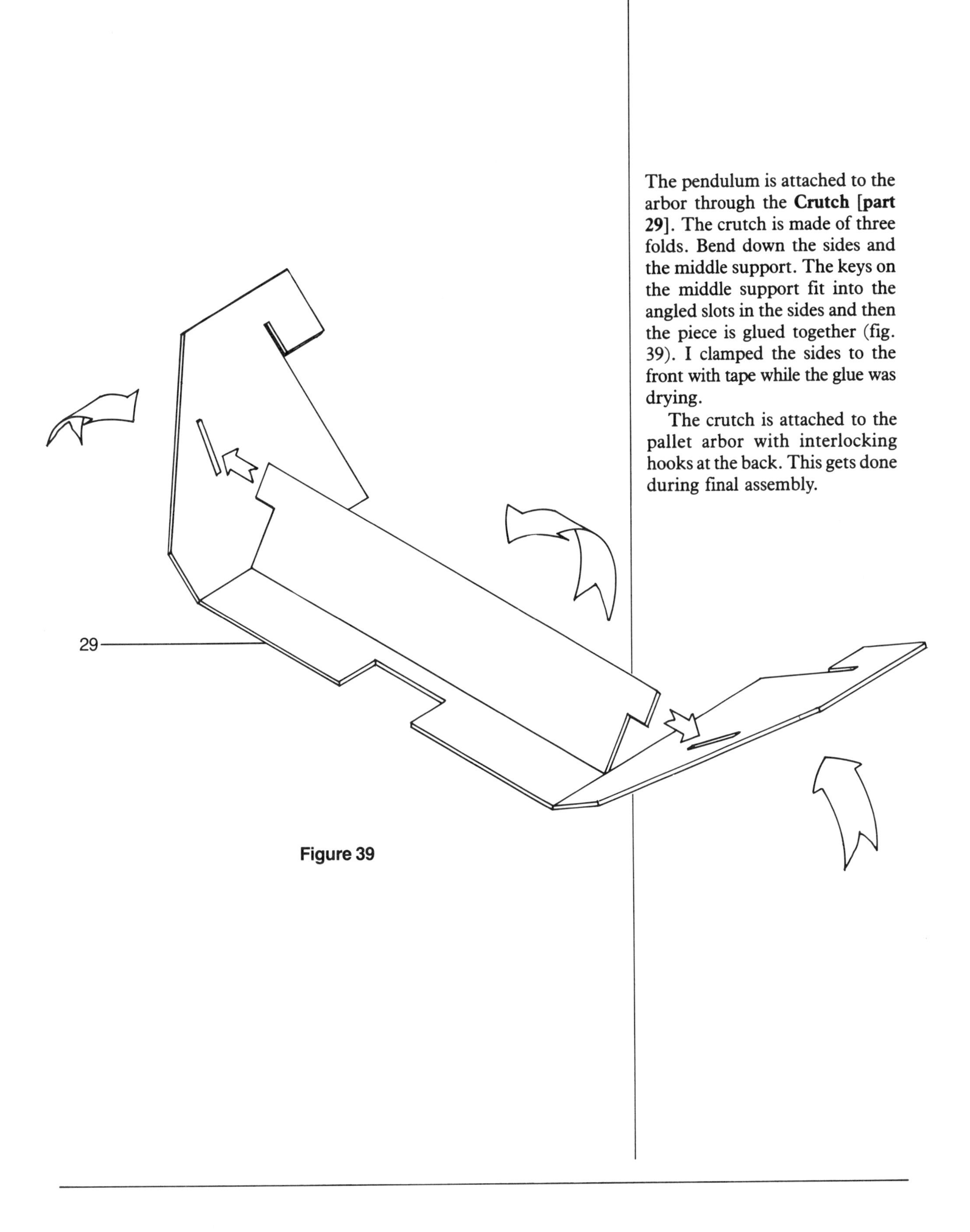

Figure 39

The pendulum is attached to the arbor through the **Crutch [part 29]**. The crutch is made of three folds. Bend down the sides and the middle support. The keys on the middle support fit into the angled slots in the sides and then the piece is glued together (fig. 39). I clamped the sides to the front with tape while the glue was drying.

The crutch is attached to the pallet arbor with interlocking hooks at the back. This gets done during final assembly.

The Frame

With the three arbors built, all that remains is to assemble the frame. The arbors turn in three pairs of pin holes that are pricked into the frames. Prick the holes with extra pins of the type used in the arbors or with a push pin. Both **Frame halves [parts 13 and 14]** have centerlines that show where the holes must be made. Place each frame on a flat surface and individually prick each hole. Hold a pin exactly over the point where the centerlines cross and press the pins right through. It is very important to make the holes in the right locations. Once the pin is mostly through you can pick up the frame and continue pushing the pin through the paper (fig. 40). I used a bunch of pins because I kept on bending them while I was forcing them through. The pivot pins should turn easily in the holes.

After the arbor holes are made, bend the sides of the frame back. The folds give the paper a tremendous amount of strength.

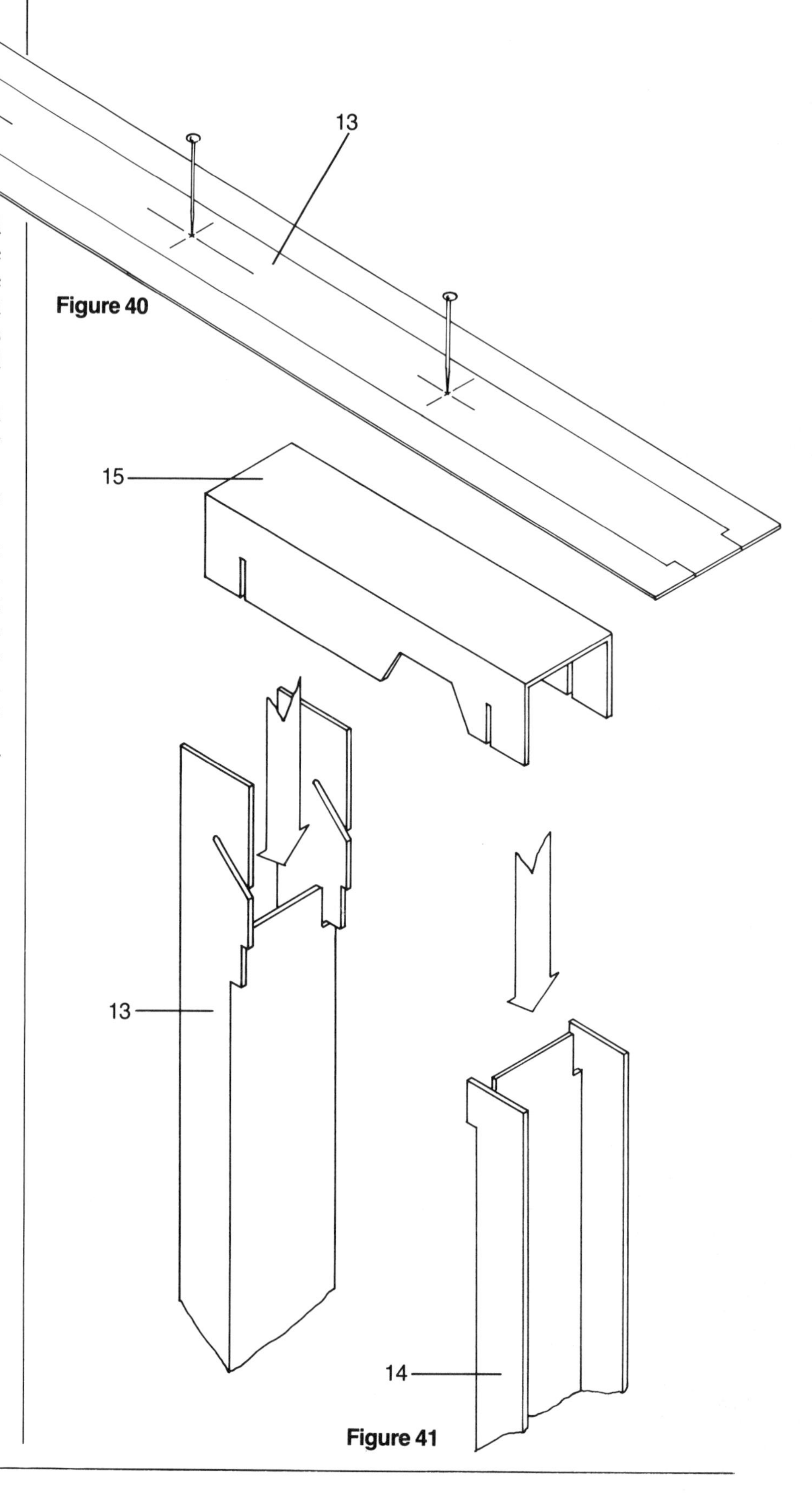

Figure 40

Figure 41

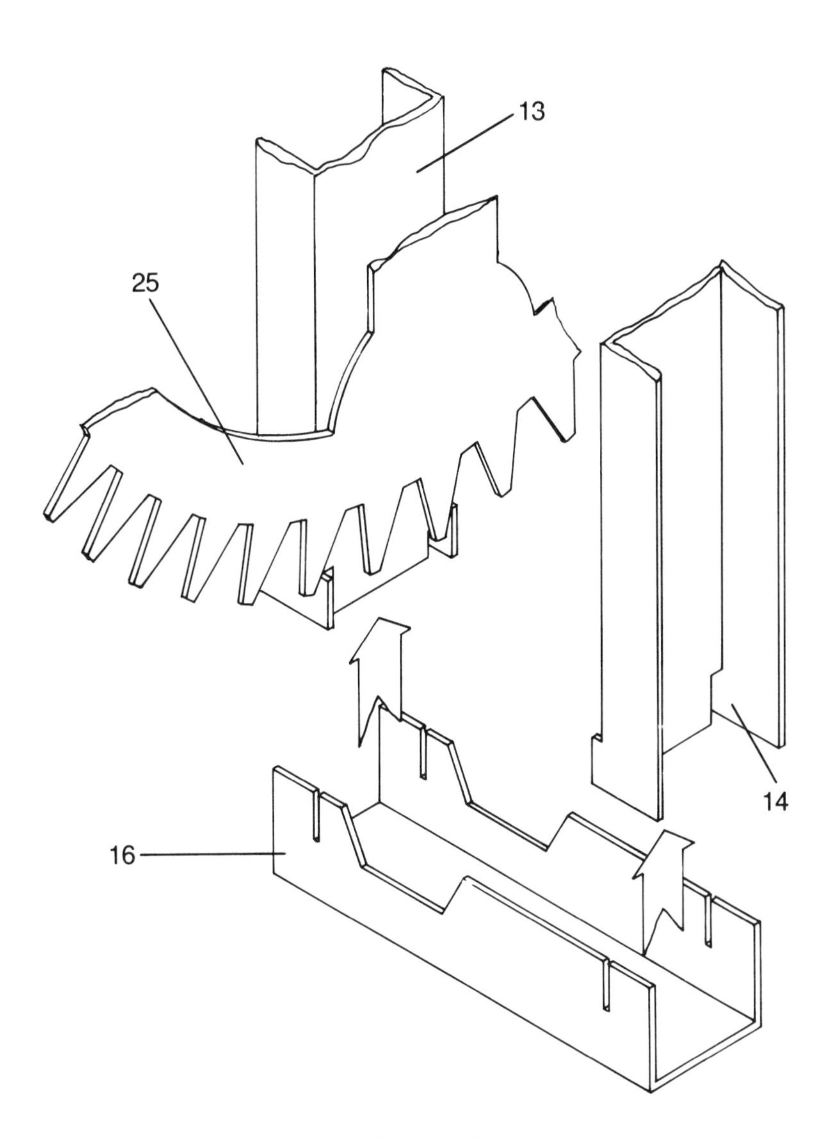

Figure 42

Bend up the **Frame Spacers [parts 15 and 16]**. These spacers are identical except for the gluing pattern for the stabilizer, which is added to the bottom spacer. The spacers slide onto the ends of the frames and interlock. For now, just test-fit the frame assembly to make sure the interlocking slots are clean. Slide the top spacer into place (fig. 41), followed by the bottom spacer (fig. 42). The cutout on the top spacer should be at the front to give clearance for the pallet arbor, and the bottom spacer cutout should be at the back so that the great wheel has room to move.

Once you are sure everything fits, assemble the frame. Clean off the points of the pivot pins by scraping them with a knife to get rid of any glue that may have accidentally gotten on them. Lay the rear frame flat up and place all three arbors in their respective bearing holes. The great wheel goes in the bottom, the escape arbor goes in the middle, and the pallet arbor fits into the hole at the top. Make sure that the great wheel engages the pinion on the escape arbor. Ease the front frame over the three pivots and then slide on the spacers to hold everything together. It's not a good idea to glue the entire frame together just yet, because you may still have to make some drastic adjustments, so just glue on the bottom spacer and tape the top one for now. That way you should be able to force the frames apart if necessary. When you tape the frame, and later when you finally glue everything together, make sure that the frame is square. **This is important.** The weights and the pendulum will try to pull everything askew when the timer is hung up and if the timer sags the gears and escapement won't mesh properly.

Once the frame is dry and taped, the timer will be rigid so you can hold it up and accurately locate the last piece. Center the **Stabilizer [part 23]** on the back of the bottom spacer. Glue it so that it is flush with the back edge of the frame (fig. 43). The stabilizer leans against the wall after the timer is hung up and steadies the frame while the pendulum swings back and forth.

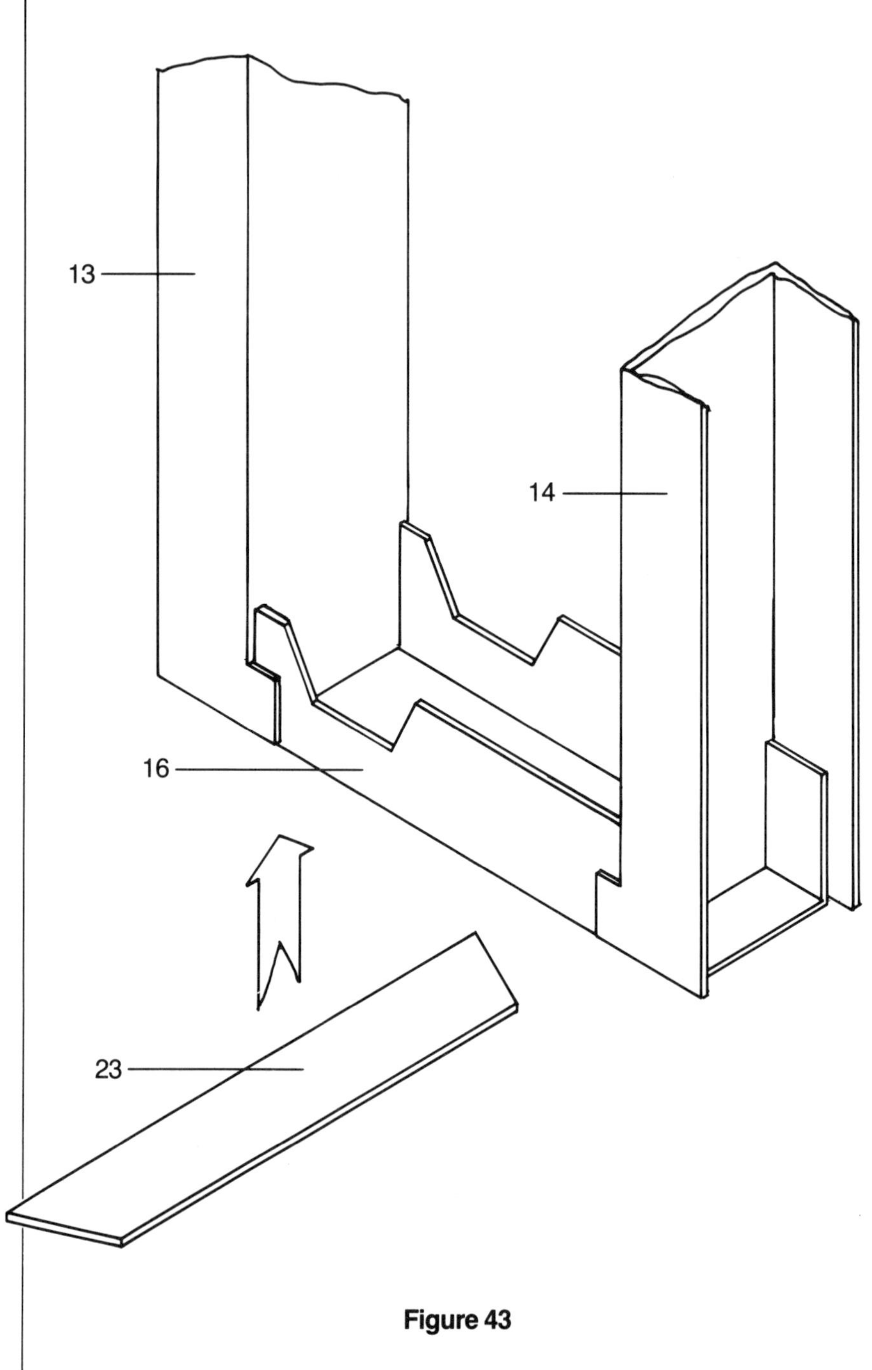

Figure 43

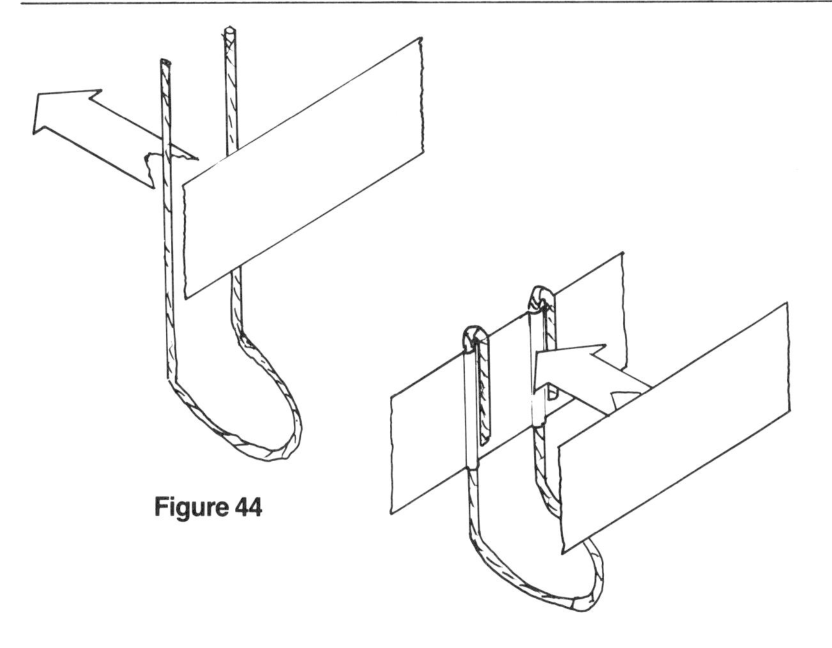
Figure 44

Figure 45

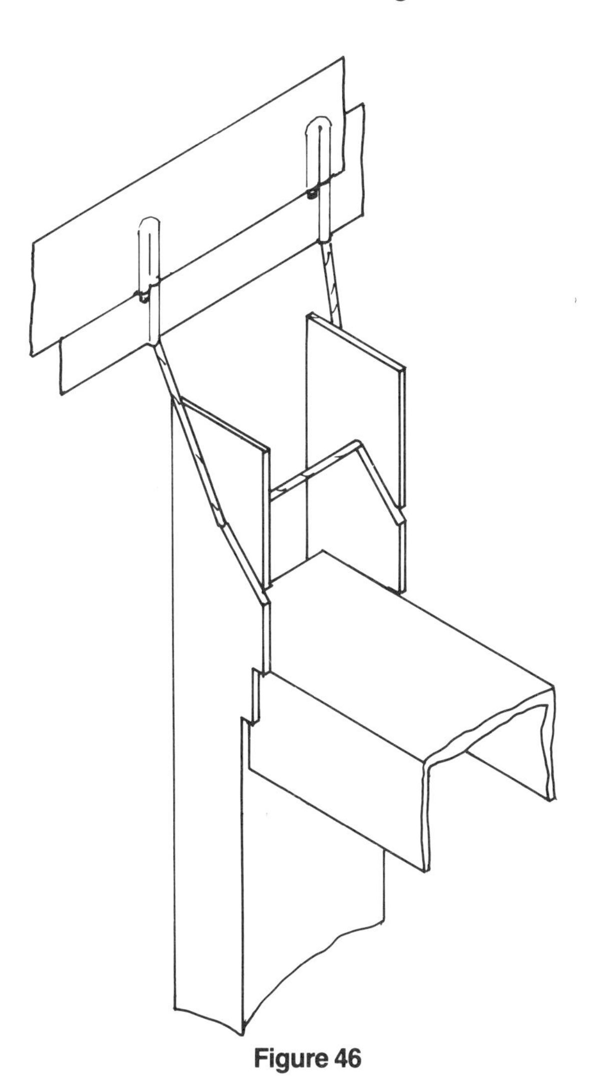
Figure 46

Final Assembly

Even after you complete the construction of the model, you will still have to make a number of adjustments before the timer will work reliably. The first thing to do is to attach the timer to a wall.

Attaching the Timer to a Wall

Pick a straight wall that has plenty of unobstructed space for the weight and pendulum. I hung my timer about five feet from the ground.

The timer is suspended from a loop made out of string that is taped to the wall. Use masking tape so that the wall isn't damaged. To make the loop, take a 6-inch piece of string and fold it in half. With the loop pointing down, tape the two loose ends to the wall with a piece of tape placed over the string about an inch from the free ends (fig. 44). Then fold the free ends back over the tape and place another piece of tape over the first piece and over the free ends, locking everything in place (fig. 45).

The timer is hung from the loop formed in the string by the two hooks in the top of the frame (fig. 46). After you hang the timer up, shift it around until the frame hangs approximately vertically. The stabilizer should be taped to the wall with a small piece of tape so that the timer doesn't shift while it is running.

Hook the crutch to the pallet arbor (fig. 47), making sure to seat it firmly in its slots. You may wish to glue or tape the crutch in place so it doesn't shift out in use.

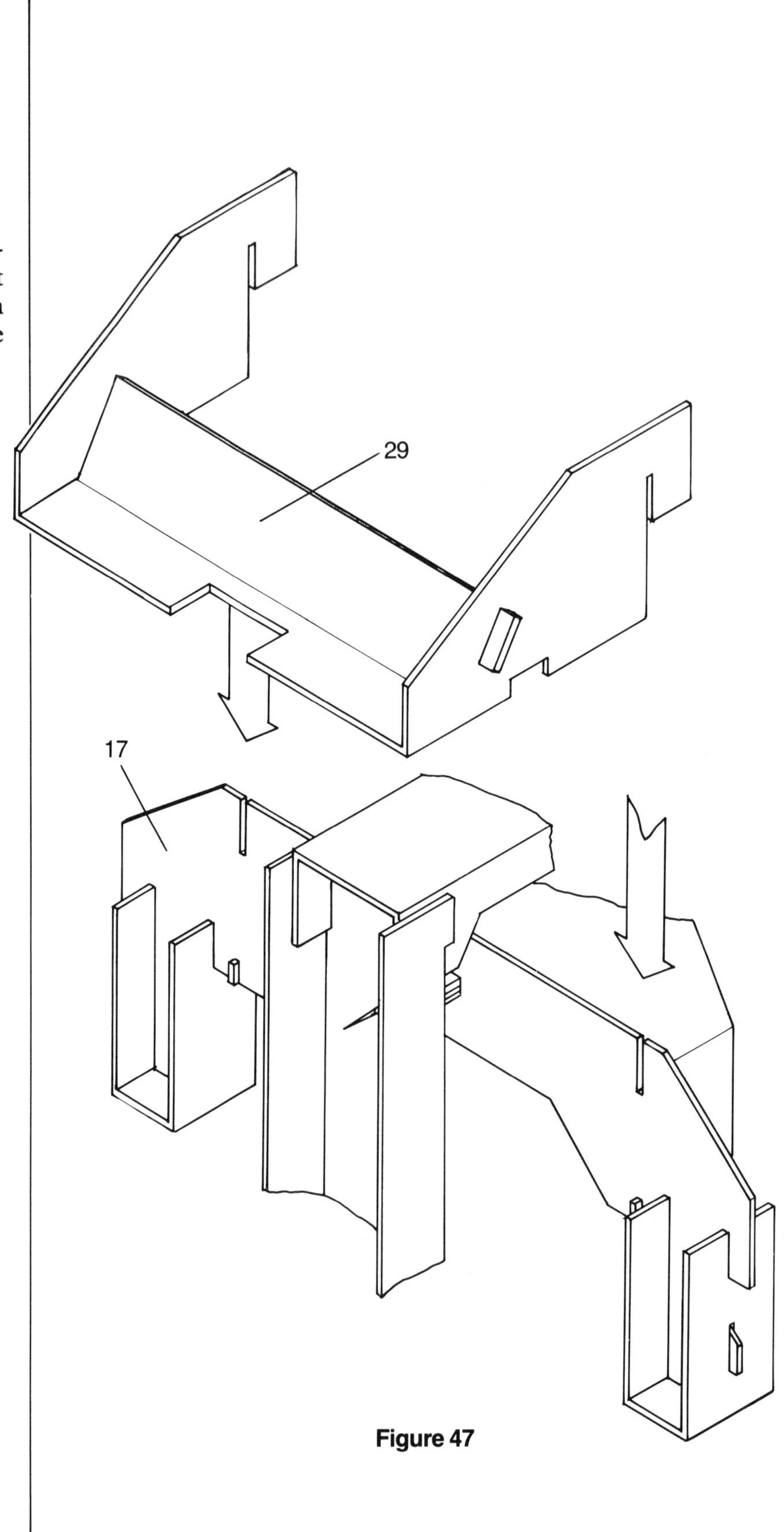

Figure 47

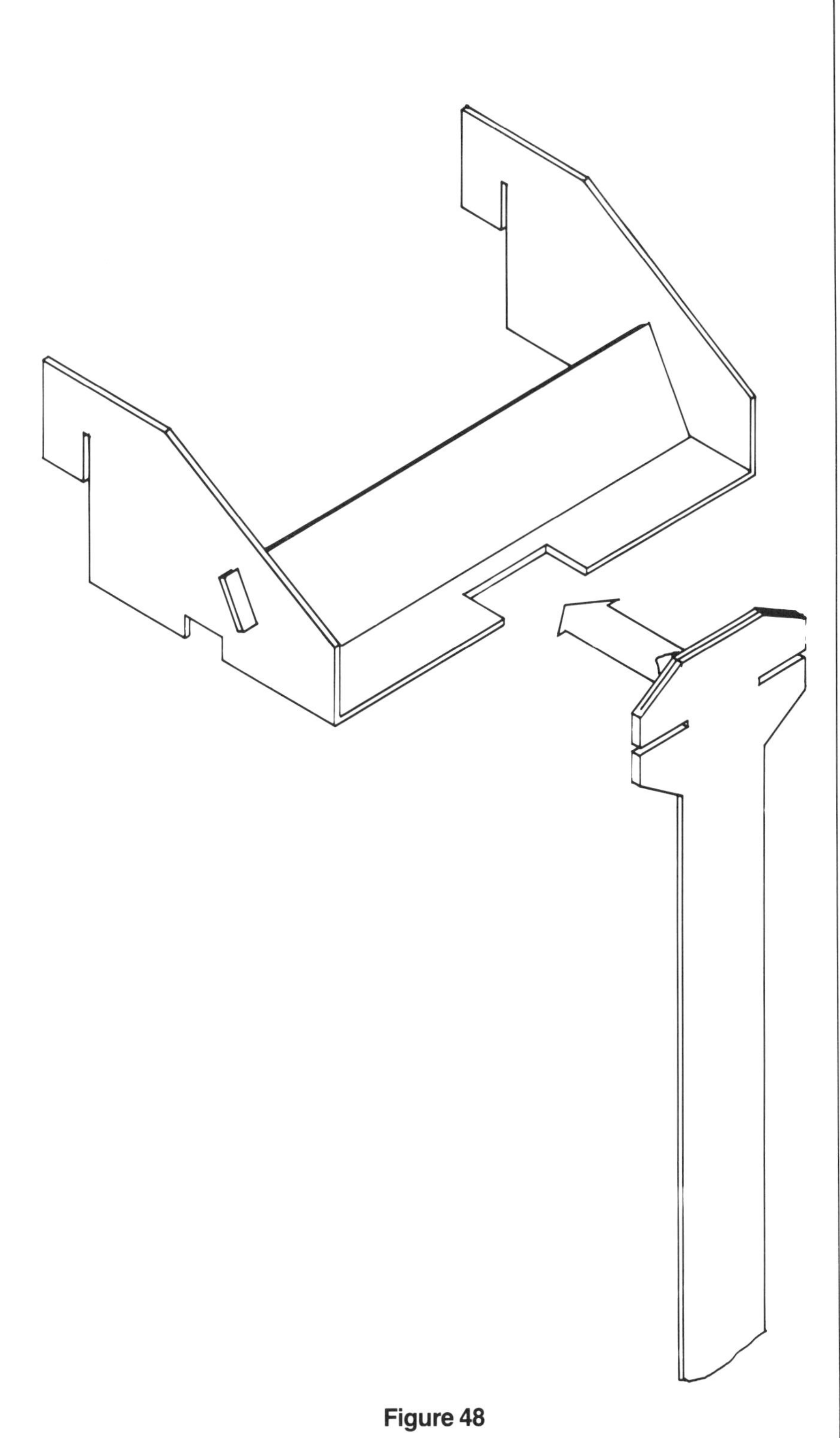

Figure 48

Slide the pendulum into the cut-out in the crutch (fig. 48). Don't glue it in. The parts will stay together but later you will still be able to disassemble the timer.

Adding the Weights

Before you attach the weight case to the power cord check both cords to make sure neither is tangled in the mechanism and that the winding cord is correctly wound on the outer drum. Fill the weight case with fifteen pennies. Stick the string tie through the hole in the top of the weight case and then straighten the tie out so that it supports the weight case at the top bend. Let the weight case hang free. Partially wind the timer by gently pulling on the winding cord. As the winding cord unwinds, the weight cord will wind around the center drum and the ratchet will make a "click-click" as the geartrain slips.

Let the weight case hang free and then give the pendulum a little sideways swing. The pallet teeth will swing back and forth and the timer will start to run. Do not be alarmed if it stops right away or if the weight case falls free; the next task is to adjust the timer so it will run continuously. This last phase of adjustment is common even to real clocks. This is where you compensate for any little errors in construction and manufacture.

Adjusting the Frame Angle

The first adjustment you will probably have to make is to get the frame in the correct orientation. Because the exact final geometry of the clock depends on many factors, the final position of the frame will not be exactly vertical. If the escape wheel stops, and always hangs up on the same side of the pallet arbor, swing the frame over a little to move the escape wheel out of the way (fig. 49). Test it again, and if the other side hangs up you may have to move the frame back a little. I taped the stabilizer to the wall with a little piece of tape so the frame wouldn't move between adjustments and while the timer was running.

If a pallet keeps skipping teeth, then move the frame back toward it.

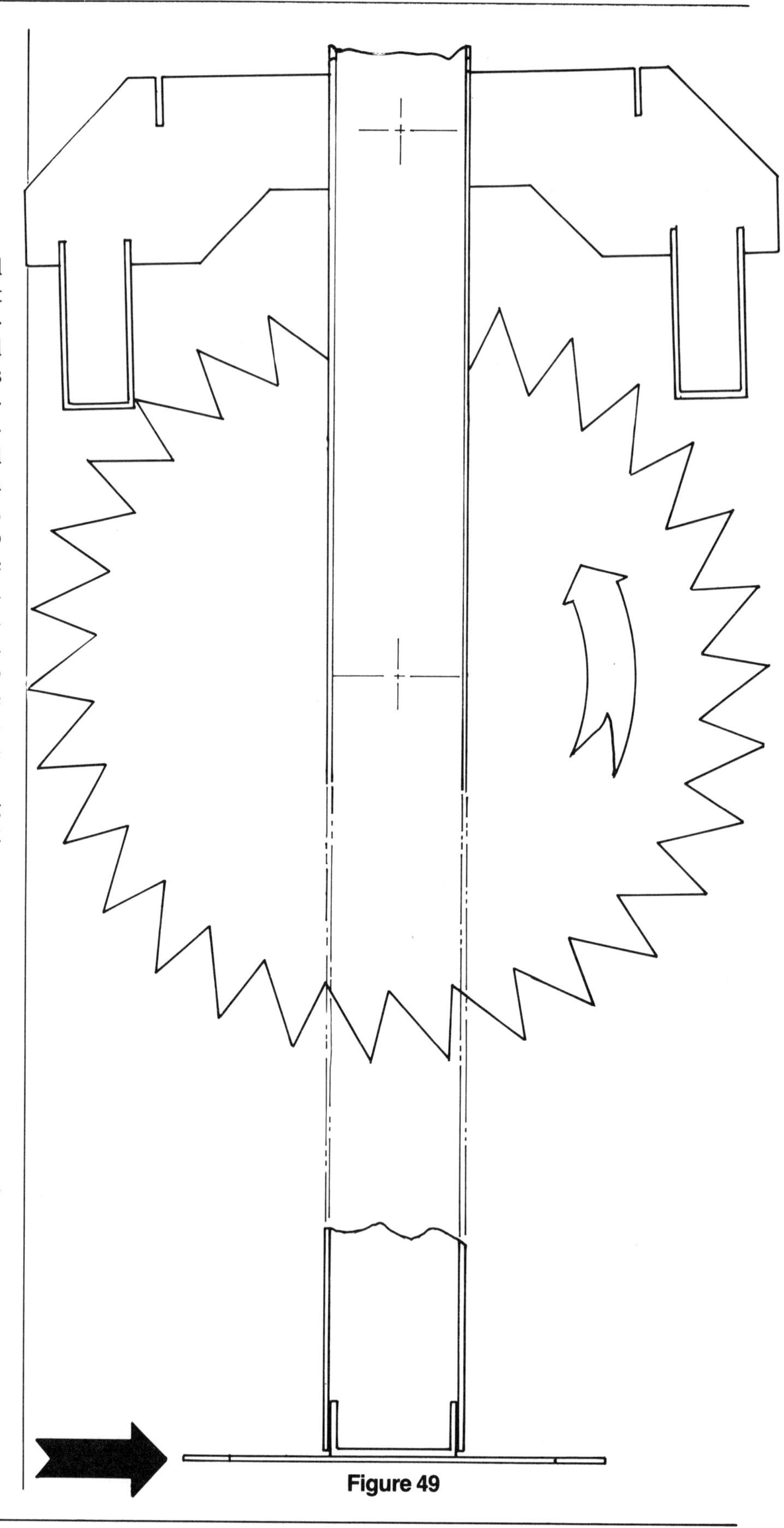

Figure 49

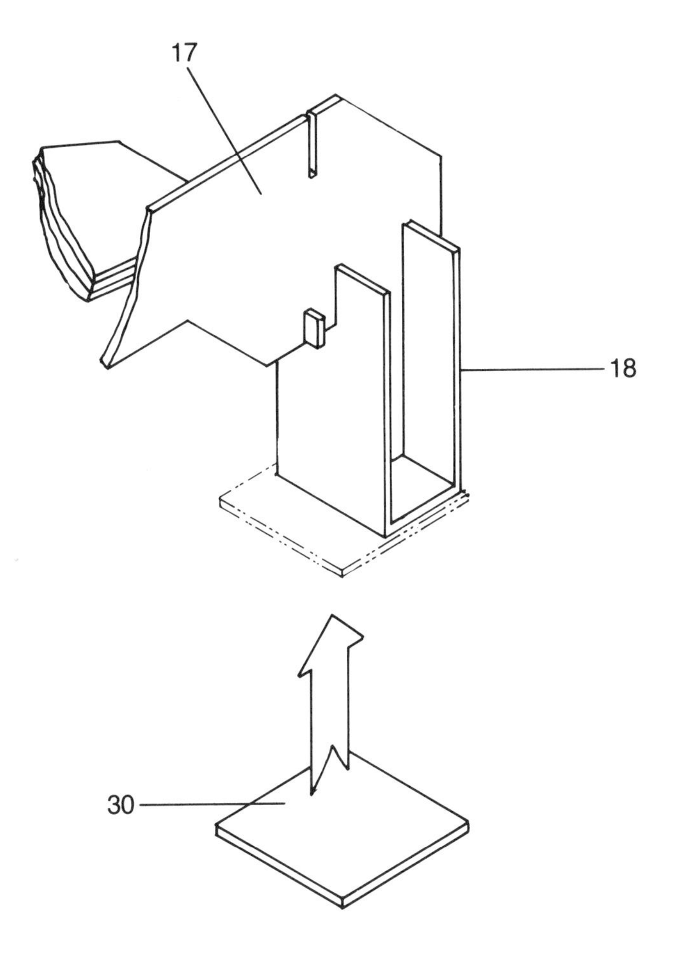

Figure 50

Shimming the Pallet Teeth

There is an excellent chance that no matter how much you adjust the frame the pallets will still skip a lot of teeth, or skim over them, and the weight case will run free for most of a revolution of the escape wheel. The solution to this problem is to glue a **Shim [part 30–1 or 30–2]** to the bottom of the righthand pallet. The kit has two shims, but you will probably need only one. Glue the shim to the pallet with about a $3/16$-inch overhang (fig. 50). After the glue dries try the timer again. Chances are that you will have to readjust the angle of the frame again. This time the clock may not run at all because the shim is too long and the escape wheel will hang up on both sides of the pallet. Trim the shim back slightly, and adjust the frame until the escape wheel turns and catches and stops each pallet in turn. It may take several trimmings to get everything right. If you find the wheels skipping again you may want to add the second shim to the left pallet. This time glue the shim to the side of the pallet. The parts drawing (see figure 1 on page 8) shows both shims in their correct locations.

If, when you first test the timer, before you glue on shims, it becomes apparent that the pallet arbor is too close to the escape wheel, and both pallets simultaneously bind, then take off the top spacer, force the frame apart to remove the arbor and make new pivot holes. This correction is a major adjustment and I never found it necessary. The only time I could see a need for it is if the pallet arbor holes were pricked too close to the escape wheel hole in the first instance, and everything works against you.

If the timer keeps stopping and it becomes clear that the pallets are not blocking the escape wheel, check the great wheel and pinion. (Also check the obvious—that the timer doesn't simply need rewinding.) If the frame is not square, the great wheel can catch on the sides of the pinion, or if either the pinion or great wheel are out of shape, there may be a rough spot that the great wheel isn't strong enough to overcome. In the former case, straighten up the frame so that the great wheel rides in the center of the pinion. In the latter case, make sure that the pinion teeth are correctly seated and that there are no stray paper scraps on the great wheel. If the timer always stops on the same tooth, it is definitely a problem with that tooth.

Sometimes, adding a few more pennies to the weight case will clear up any stoppage, as will lubricating the arbor pivots by greasing the pivot pins with a little Vaseline or some other grease. Just apply a little, on the inside of the frames, where the arbor pivot holes are.

The only adjustments I had to do to my models were adjusting the angle of the frame and adding a shim to the bottom of the righthand pallet.

Setting the Speed

With the adjustments finished, and the timer working smoothly, the model is completed and the only thing left that you may want to do is to get it to keep exact time. My goal was to set the timer so that it would take exactly four minutes to complete a single revolution of the great wheel. That's why the four spokes on the great wheel are numbered from one to four. Each quarter represents a single minute. You could ignore my markings and make the timer go much faster by shortening the pendulum, or even removing it. I suppose, though I've never tried this, you could hang the timer over a stairwell and get a much longer string and then see how many hours you can get the timer to run continuously. Or add a longer pendulum. Personally, I sometimes get transfixed and just watch the pendulum going back and forth, back and forth, tick-tock, tick-tock. . . .

A SHORT HISTORY OF THE PENDULUM CLOCK

Timepieces of many different forms and constructions have been known since ancient times. The earliest clocks were forms of sundials, such as Stonehenge in Wiltshire, England, which is roughly 4000 years old. It kept track of the seasons.

The ancient Greeks and Romans built water clocks and geared astrolabes, a sort of mechanical calendar. The first weight-driven mechanical clocks were used to ring bells to signal important intervals during the day. The actual date of the invention is lost but sometime between 1200 and 1300 A.D. primitive mechanisms with a single hour hand started to appear in church towers and in marketplaces across Europe. The invention of a device that could ring a bell was so revolutionary that a new word was coined to describe the machine. Instead of calling it a *horologium*, the Latin for "water clock" (and the root of a word that is used today to describe the whole world of clockmaking, "horology"), the word "clock" was used. Clock is derived from the Latin *clocca*, which means "bell."

The Church was a primary customer for the first generation of clocks because prayers were said at specific times during the day and sundials could not always be counted upon.

There were two major problems with the first generation of mechanical clocks: Europe was just recovering from the chaos caused by the collapse of the Roman Empire and the technical skill required to produce an accurate set of gears that could run continuously did not exist. Secondly, the escapements, the devices used to control the speed of the clock, were not accurate. The accuracy of a clock depends on an escapement beating at absolutely regular intervals, and before 1650 even a perfectly made escapement wouldn't keep very accurate time; the mechanisms had built-in design limitations. The clocks didn't keep particularly good time; in fact, most had only a single hour hand because they weren't accurate enough to have a meaningful minute hand. This lack of accuracy restricted the usefulness of clocks in science and navigation.

Clocks were more of a status symbol than a practical instrument. Just as today, where national pride is one justification of the space program, towns and villages of the Middle Ages competed to see who could buy the biggest and best clock. Clockmakers responded to the demand for fancy mechanisms and added music boxes and animated figures, or automata, to a basically primitive timekeeper. This was the start of the great age of performing clocks: clocks that had marching soldiers, dancing animals, orchestras, and other moving sculpture all jointed and linked to a mechanism that would animate the entire system at specific intervals. Centuries later, in the 1800's, during the industrial revolution, the basic linkages that controlled these curiosities were put to practical application to control weaving looms. Those looms were the forerunners of the first computers and in a very real sense the idea of "intelligent" machines can be said to date back six centuries.

The effect a fancy tower clock must have had on a public that had no television, no electricity, and few mechanical devices must have been amazing. To put this in modern perspective, the only performing clock that I have seen in America is the Delacorte Clock, which is in the Central Park Zoo, next to the Arsenal, in New York City. It's a recent addition to the park, built in 1965 and much simpler than the fancy devices of the Middle Ages that were its inspiration. On the half hour, bronze statues of animals dance around the clock tower and one of thirty-two tunes is loudly played. The repertory includes nursery rhymes such as "Hickory dickory dock, the mouse ran up the clock." What I find amazing is that even in today's complicated technological society, people of all ages gather around to hear the clock play. Passersby sometimes stop for a few minutes if it is near the hour, and everyone who is in the area stops if it's ringing, just to hear it go.

In the Middle Ages such an exciting new gadget was bound to get people thinking. For the first time, people saw a device that took on a life of its own. Some people thought that if a clock could move on its own, without the aid of gods to move gears and bells, maybe the whole universe was a giant clock that moved by itself, following regular laws of motion. God may have created the world, but the day to day operation was guided by natural laws that humans could discover. Instead of arbitrary motion based on fate and the whims of the gods, everyone and everything was part of a "clockwork universe."

The average person in the Middle Ages didn't need a clock. People worked from sunrise to sunset, and there was no such thing as a minute by minute schedule. However, there was a distinct need for accurate timekeeping in the fields of physics and navigation. It is appropriate that the next step in technology wasn't made by a clockmaker but a scientist.

According to legend, in 1582 Galileo Galilei, the famous astronomer, was sitting in church and noticed that the chandelier was swinging back and forth in a breeze. By timing the swings with his pulse he was able to determine that the duration of each swing was constant, no matter if the angle of swing was long or short. Later, he discovered that the duration was constant for each length of pendulum. Galileo had discovered an accurate method of keeping time that he could use in his physics experiments. All he had to do was hang up a pendulum, let it swing, and count the number of swings it made. The only drawback was that periodically he had to give the pendulum a push to keep it swinging.

The next step was inevitable. Privately, in about 1641, he had his son sketch for him the first pendulum escapement. The drawing shows a mechanism very similar to existing movements of the day, but the inaccurate method of timing the escape wheel was replaced by a pendulum. The pendulum controlled the speed of the escape wheel, and the escape wheel imparted energy to the pendulum to keep it going. The mechanism was never built, and the drawing was temporarily lost after his death.

Christiaan Huygens, the mathematician, physicist, and astronomer, was faced with the same problem about time. His solution was identical. Unlike Galileo, he had a prototype built and announced his invention immediately in 1656, only to be instantly accused of plagiarizing Galileo. The effect of Huygens's invention was stupendous. Clock accuracy improved by several orders of magnitude, and it became possible to add minute and second hands to clocks.

The primary difference between Huygens's design and Galileo's was that Huygens isolated his pendulum and connected it to the pallet arbor via a separate follower piece, so that the pendulum swung free for most of its cycle. Galileo connected the pendulum directly to the pallet arbor. My design follows Galileo's because I found the added friction and complexity of a paper follower piece ate up any possible advantage the added accuracy had. However, Huygens's design was potentially more accurate, and his method was universally adopted.

After the invention of the pendulum escapement, for all intents and purposes the basic clock mechanism remained the same until this century. Minor improvements were made and better escapements were designed, but most used the pendulum and were incremental rather that revolutionary changes. Other means of powering clocks were invented instead of weights, such as coil springs. Watches that used an oscillating spring, instead of a pendulum, to regulate the time independent of gravity, made a portable timepiece possible.

The 1800s—The Industrial Revolution

Almost from the beginning, clocks were made in a kind of system of mass production. Masters, owners of clockshops, would contract the bulk of manufacture to skilled and semiskilled workers who would either work at home, in their own shop, or even in the same shop as the master, and each would produce a single group of parts for a clock. There were gearmakers, springmakers, dialmakers, casemakers, etc. Clocks were produced in batches and, while the parts were by no means interchangeable, each batch was made to the same design. After the master had assembled all the parts from the various job shops, his staff would custom-fit and assemble the final product. The work was done entirely by hand. Gears would be laid out and then filed, by hand. Housings were hand-built by casemakers who either worked as subcontractors or bought unfinished movements, mounted them in cases, and sold them. The primitive machine tools that did exist were aids to human skill, not a substitute for it. The semiskilled workers who produced the clock parts could not produce an entire movement, but they were very skilled in their little area of specialization.

The early days of the industrial revolution intensified this system. Better tools were invented and materials became cheaper. Many simplifications to the basic clock movement were made to reduce costs but the basic hand manufacturing techniques remained.

While this was happening in Europe, the newly formed United States was merely an underdeveloped, former colonial backwater market for the Europeans. When the Americans began to industrialize, they did so with a vengeance. The traditions and market segmentation that kept the European companies in order didn't exist here, and when a native industrial base was built it was built on a new framework, with none of the old ties to Europe.

In 1808, Eli Terry introduced the factory system to clock-making and brought the price down to a level where anyone could afford a clock.

Eli Terry did two things. First, he grouped everybody under one roof and, instead of using independent contractors, everyone became a factory employee. Second, he designed a very simple clock and then built a whole range of special machines and fixtures to make the clock so that there would be very little final fitting. His first clocks were all made out of wood, because wood was cheap.

His methods produced a product that cost less than anything produced up to that time, and he sold all he could make. As clocks became more and more a mass market item, the skills of a master clockmaker, who could build an entire timepiece by hand, disappeared.

Today, the mass market pendulum clock has been superseded by the quartz electric clock. Since customers still enjoy the magic of moving parts, many quarts movements have fake pendulums that aimlessly swing back and forth. Pendulum clocks are still made, but they are not really a mass item; their customers are people who enjoy real machinery and real chimes.

There are many collections of clocks in this country, and many museums still keep a few of the old clocks working so that visitors can see them go.

Clock collecting has always been a popular hobby, and many hobbyists even build their own clocks. I am aware of only one other published design for a paper clock. A French publication, dating from before World War II, was available in this country until at least 1978. Several plastic clock kits are available and these kits are well within the ability of anybody who has built this model.

For people who have access to metalworking equipment building clocks has always been a popular diversion. Many hobbyists without metalworking tools have made fabulous clocks out of wood. It is really a question of enthusiasm rather than equipment.

I hope you had a good bit of fun putting the model together. If you enjoyed building this model you may also enjoy building its companion in the Paper Machine Series—*The Working Piston Engine*, which will be available soon and which contains all the precut paper parts you will need to make a working model of a steam engine powered by a small balloon.

If you are interested in finding out more about model-making, the history of technology, or if you have any questions about this book, feel free to drop me a line at

Joel Moskowitz
c/o Simon and Schuster
1230 Avenue of the Americas
New York, New York 10020
Attn: Paper Machine Series

and I'll do my best to point you in the right direction.

ABOUT THE AUTHOR

Joel Moskowitz is an engineer who lives in New York City.